AF731992

Regulatory Governance and Risk Management

Routledge Advances in Management and Business Studies

For a full list of titles in this series, please visit www.routledge.com

12 **Telecommunications Regulation**
Culture, Chaos and Interdependence Inside the Regulatory Process
Clare Hall, Colin Scott and Christopher Hood

13 **Rethinking Public Relations**
The Spin and the Substance
Kevin Moloney

14 **Organisational Learning in the Automotive Sector**
Penny West

15 **Marketing, Morality and the Natural Environment**
Andrew Crane

16 **The Management of Intangibles**
The Organization's Most Valuable Assets
A Bounfour

17 **Strategy Talk**
A Critique of the Discourse of Strategic Management
Pete Thomas

18 **Power and Influence in the Boardroom**
James Kelly and John Gennard

19 **Public Private Partnerships**
Theory and Practice in International Perspective
Stephen Osborne

20 **Work and Unseen Chronic Illness**
Silent Voices
Margaret Vickers

21 **Measuring Business Excellence**
Gopal K Kanji

22 **Innovation as Strategic Reflexivity**
Edited by Jon Sundbo and Lars Fuglsang

23 **The Foundations of Management Knowledge**
Edited by Paul Jeffcutt

24 **Gender and the Public Sector**
Professionals and Managerial Change
Edited by Jim Barry, Mike Dent and Maggie O'Neill

25 **Managing Technological Development**
Hakan Hakansson and Alexandra Waluszewski

26 **Human Resource Management and Occupational Health and Safety**
Carol Boyd

27 **Business, Government and Sustainable Development**
Gerard Keijzers

28 **Strategic Management and Online Selling**
Creating Competitive Advantage with Intangible Web Goods
Susanne Royer

29 **Female Entrepreneurship**
Implications for Education, Training and Policy
Edited by Nancy M. Carter, Colette Henry, Barra Ó Cinnéide and Kate Johnston

30 **Managerial Competence within the Hospitality and Tourism Service Industries**
Global Cultural Contextual Analysis
John Saee

31 **Innovation Diffusion in the New Economy**
The Tacit Component
Barbara Jones and Bob Miller

32 **Technological Communities and Networks**
International, National and Regional Perspectives
Dimitris G. Assimakopoulos

33 **Narrating the Management Guru**
In Search of Tom Peters
David Collins

34 **Development on the Ground**
Clusters, Networks and Regions in Emerging Economies
Edited by Allen J. Scott and Gioacchino Garofoli

35 **Reconfiguring Public Relations**
Ecology, Equity, and Enterprise
David McKie and Debashish Munshi

36 **The Pricing and Revenue Management of Services**
A Strategic Approach
Irene C. L. Ng

37 **Critical Representations of Work and Organization in Popular Culture**
Carl Rhodes and Robert Westwood

38 **Intellectual Capital and Knowledge Management**
Strategic Management of Knowledge Resources
Federica Ricceri

39 **Flagship Marketing**
Concepts and Places
Edited by Tony Kent and Reva Brown

40 **Managing Project Ending**
Virpi Havila and Asta Salmi

41 **AIDS and Business**
Saskia Faulk and Jean-Claude Usunier

42 **The Evaluation of Transportation Investment Projects**
Joseph Berechman

43 **Urban Regeneration Management**
International Perspectives
Edited by John Diamond, Joyce Liddle, Alan Southern and Philip Osei

44 **Global Advertising, Attitudes, and Audiences**
Tony Wilson

45 **Challenges and Controversies in Management Research**
Edited by Catherine Cassell and Bill Lee

46 **Economy, Work, and Education**
Critical Connections
Catherine Casey

47 **Regulatory Governance and Risk Management**
Occupational Health and Safety in the Coal Mining Industry
Binglin Yang

Regulatory Governance and Risk Management

Occupational Health and Safety in the Coal Mining Industry

Binglin Yang

Routledge
Taylor & Francis Group
NEW YORK AND LONDON

First published 2012
by Routledge
605 Third Avenue, New York, NY 10017

Simultaneously published in the UK
by Routledge
4 Park Square, Milton Park, Abingdon, Oxon OX14 4RN

Routledge is an imprint of the Taylor & Francis Group, an informa business

© 2012 Taylor & Francis

The right of Binglin Yang to be identified as author of this work has been asserted by him in accordance with sections 77 and 78 of the Copyright, Designs and Patents Act 1988.

Typeset in Sabon by IBT Global.

All rights reserved. No part of this book may be reprinted or reproduced or utilised in any form or by any electronic, mechanical, or other means, now known or hereafter invented, including photocopying and recording, or in any information storage or retrieval system, without permission in writing from the publishers.

Trademark Notice: Product or corporate names may be trademarks or registered trademarks, and are used only for identification and explanation without intent to infringe.

Library of Congress Cataloging-in-Publication Data

Yang, Binglin, Ph. D.
Regulatory governance and risk management : occupational health and safety in the coal mining industry / by Binglin Yang.
p. cm. — (Routledge advances in management and business studies ; 47)
Includes bibliographical references and index.
1. Coal mines and mining—United States—Safety measures. 2. Coal mines and mining—Risk assessment—United States. 3. Mineral industries—Risk management—United States. 4. Mineral industries—United States—Safety measures. 5. Mine safety—Law and legislation—United States. 6. Mine safety—Law and legislation—Australia. I. Title.
HD7269.M62U692 2011
363.11'96223340973—dc22
2011009756

ISBN13: 978-0-415-89712-9 (hbk)
ISBN13: 978-0-203-80438-4 (ebk)

For my parents

Contents

Figures

Tables

Abbreviations

ALARP	As Low As Reasonably Practicable
BBS	Behavior-Based Safety
BST	Behavioral Science Technology, Inc.
BTA	Bow Tie Analysis
CCP	Cooperative Compliance Program
CFMEU	Construction Forestry Mining Energy Union
CMHSA	Coal Mine Health and Safety Act, 2002 (NSW)
CMSHA	Coal Mining Safety and Health Act, 1999 (Qld)
Coal Act	Coal Mine Health and Safety Act, 1969
DOL	Department of Labor
EIA	Energy Information Administration
EMS	Environmental Management System
EPA	Environmental Protection Agency
FIFR	Fatal Injury Frequency Rate
FTA	Fault Tree Analysis
GAO	Government Accountability Office
GMI	Global Mining Initiative
HACCP	Hazards Analysis and Critical Control Points
ICMM	International Council on Mining & Metals
IIED	International Institute for Environment and Development
ILO	International Labor Organization
INPO	Institute of Nuclear Power Operations
JSA	Job Safety Analysis
LTI	Lost-Time Injury
MCA	Minerals Council of Australia
MHRA	Major Hazard Risk Assessment
MHSA	Mine Health and Safety Act, 2004 (NSW)
Mine Act	Federal Mine Safety and Health Amendment Act, 1977
Miner Act	Mine Improvement and New Emergency Response Act
MIRM	Minerals Industry Risk Management
MISHC	Minerals Industry Safety and Health Centre
MMSD	Mining, Minerals and Sustainable Development
MNM	Metal and Non-Metal

MQSHA	Mining and Quarrying Safety and Health Act, 1999 (Qld)
MSAC	Mine Safety Advisory Council
MSHA	Mine Safety and Health Administration
NFDL	Nonfatal Days Lost
NIOSH	National Institute for Occupational Safety and Health
NMA	National Mining Association
NSW	New South Wales
OECD	Organization of Economic Cooperation and Development
OHS	Occupational Health and Safety
OHSMS	Occupational Health and Safety Management System
OPM	Office of Personnel Management
OSHA	Occupational Safety and Health Administration
PHA	Preliminary Hazard Analysis
Qld	Queensland
SLAM	Stop-Look-Analyze-Manage
S-Miner Act	Supplementary Miner Act
TMI	Three Mile Island
TQM	Total Quality Management
UMSC	Utah Mine Safety Commission
UMWA	United Mine Workers of America
VCP	Voluntary Compliance Partnership
VPP	Voluntary Protection Partnership
WBCSD	World Business Council for Sustainable Development
WRAC	Workplace Risk Assessment and Control

Preface

This book is based on my PhD dissertation at the Center for Public Administration and Policy at Virginia Tech. What motivated me to conduct this research was the growing attention on mining safety not only in developing countries, such as China, but also in developed economies, particularly the U.S. By examining the diffusion of risk-based governance in the U.S. and Australia, I hope this research can contribute to regulatory studies in terms of understanding the mechanisms that both drive and block the diffusion of regulatory practices. In addition, I would be very delighted if practitioners can find this book helpful in removing barriers to a safer mining industry.

I would like to give my special thanks to all committee members who helped me complete this research. Dr Joseph Rees, Dr Karen Hult, and Dr Larkin Dudley, I highly appreciate your tremendous influence on my intellectual development along the path of regulatory governance, your detailed guidance on the research design and even copy editing, and also all the support you provided for my life in Blacksburg, which I always consider as my hometown in the U.S. I am also thankful to Dr R. Larry Grayson and Mr. David Lauriski for your willingness to join my committee, even though we did not even meet one another when I asked. Your help in accessing companies, along with the comments and information you provided, were extremely useful to my research.

Moreover, some people, although they were not directly involved in my dissertation research, shall be recognized because of their crucial influence in my PhD study. I wish to thank CPAP and Susan Willis-Walton at the Center for Survey Research for providing me with financial assistance, as well as Dr John Rohr for his life-long influence on me that goes beyond academics.

In addition, I am grateful to all the interviewees. Many of you talked with me for several hours even when it was inconvenient. This research could not have been completed without your participation. I would also like to thank Jim Joy for reviewing this work. Finally, I deeply appreciate my parents for your unconditional support and patience. This is a gift for which you have waited too long.

Binglin Yang
Denver, Colorado

1 Introduction

DIFFERENT REGULATORY APPROACH, DIFFERENT RESULTS

At 3:27 p.m., April 5, 2010, an explosion occurred in Upper Big Branch coal mine in West Virginia, the U.S. The disaster killed 29 miners and seriously injured another two. This is the worst coal mining accident in the U.S. since 1970, when 38 miners were killed in the Finley Coal Company mine explosion. The investigation of the disaster is still undergoing. However, a lot of information shows that the explosion probably resulted from the poor maintenance of ventilation systems in the mine, because a total of 76 orders, which are related to the operator's failure to comply with the approved ventilation plan, had been issued to the mine since 2009 before the explosion occurred (MSHA, 2010a).

Unfortunately, such a catastrophic event is not the only one in recent years. Just four years ago, on January 2, 2006, a similar explosion, which occurred in the Sago mine in West Virginia, caused 12 fatalities. As a response to the Sago disaster, the U.S. Congress and the federal regulatory agency Mine Safety and Health Administration (MSHA) put much greater emphasis on coal mining safety. However, their "shoot from the hip" approach is questionable. Mine safety regulation in the U.S., which is characterized by a highly prescriptive approach emphasizing tough enforcement and penalties, has become increasingly prescriptive after the Sago accident. MSHA would increase its budget to a historical level, $357 million in 2010 (DOL, 2011). This is close to MSHA's peak budget in 1980, which was inflation-adjusted to $375 million in 2009 dollars (Figure 1.1).[1] Total citations and orders issued reached 175,079 in 2009, the highest level since 1990 and 37% more than that in 2005 (MSHA, 2010b). Six months after the disaster, the Congress enacted the Mine Improvement and New Emergency Response Act (the Miner Act), which dramatically raised civil penalties for violating the law. As a result, the amount of assessed penalties for violations in 2009 was 6 times that in 2005 (ibid). The question then is why this enhanced level of enforcement and spending has not prevented catastrophic accidents from occurring. Why have other countries, such as Australia, which have a different regulatory approach, achieved better mine safety performance than the U.S.?

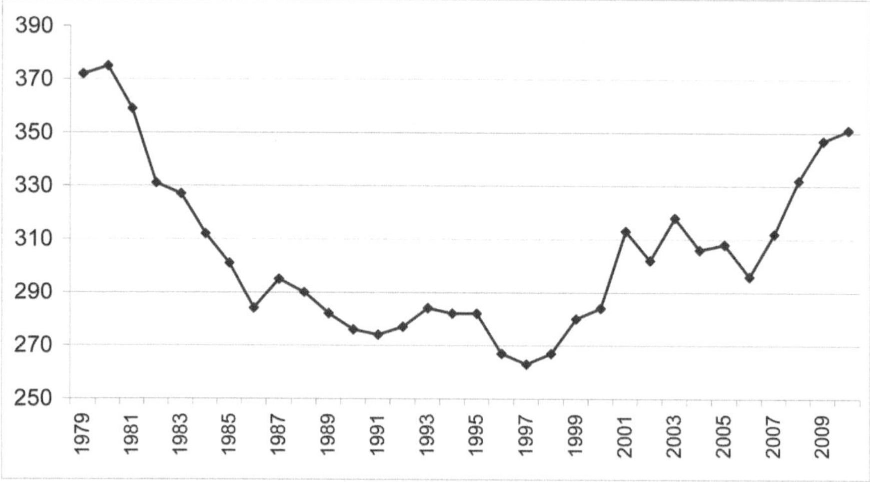

Figure 1.1 MSHA budget, 1979–2010 (millions).
Source: Budget of the U.S. Government, FY1981–2011.

Recently, there has been a rapid growth of a new regulatory approach, which I call risk-based governance in this book, in coal mining safety regulation around the world. This development of risk-based governance involves two different but interrelated processes, the growth of risk-based regulation in governments as an alternative to traditional prescriptive regulation and the increasing adoption of voluntary risk management systems in corporations. In terms of government regulation, coal mining safety regulation in most countries has long been dominated by the prescriptive approach, partly because of its distinct history and highly hazardous working environment. Mining was also covered by legislation separate from general occupational health and safety (OHS) before the 1970s. This situation gradually changed in European countries after the publication of the Robens Report in the U.K. in 1972. This report criticized traditional prescriptive regulation and emphasized that the responsibility for controlling workplace hazards should be put on those who create the hazards. Based on the recommendations of the Robens Report, the U.K. Health and Safety at Work Act was enacted in 1974. This act imposed a general duty on operators to provide safe and healthy workplaces with few prescriptive requirements. The requirement of a general duty of care implies that employers need to develop risk-based management systems to fulfill their responsibilities. This act also integrated mining regulation into general OHS regulations.

During the 1980s and early 1990s, however, when risk-based regulation spread to general OHS regulation in almost all developed countries, the mining industry still lagged behind because almost all major mining countries remained under a prescriptive regulatory regime. This changed in late 1990s when coal mining safety regulation in Australia—the largest coal exporter in the world—moved toward the approach of risk-based regulation. Queensland (Qld) and New South Wales (NSW) passed new mining statutes in 1999 and 2002, respectively. Following a Robens-style regulation, these new mining statutes embraced the general duties and systems-based standards based on risk management instead of specific standards.

Meanwhile, some of the largest mining multinationals tried to promote risk management in the mining industry worldwide (see Chapter 3 for the detailed process). Many of these leading players are Australian-based companies, such as Xtrata, Rio Tinto, and BHP Billiton. Along with the influence of mandatory requirements of new legislation, risk management systems have been widely adopted by Australian mining companies.

The change of approach in Australia has had a dramatic impact on safety performance. According to the *Safety and Health Performance Report, 2006–2007*, published by the Minerals Council of Australia, from 1997 to 2007, the Australian minerals industry average fatal injury frequency rate (FIFR) was 0.07, compared with the U.S. rate of 0.17. For the coal sector, FIFR in Australia was 0.06, which is substantially lower than the American equivalent rate of 0.16 (Minerals Council of Australia, 2007). By contrast, in the period 1989–1999, Australia had just slightly better average FIFR than the U.S. in coal mining and performed worse in metal mining (Minerals Council of Australia, 1999). Figure 1.2 shows that Australia has gone through a much more remarkable improvement in safety performance than the U.S. during the period of 1991–2007.

There is no doubt that other factors may play a role in the improving outcomes achieved in Australia. For example, since the 1990s, there has been a much larger scale of consolidation in Australia. Consequently, fewer small mining companies operate in Australia than in the U.S. Since many

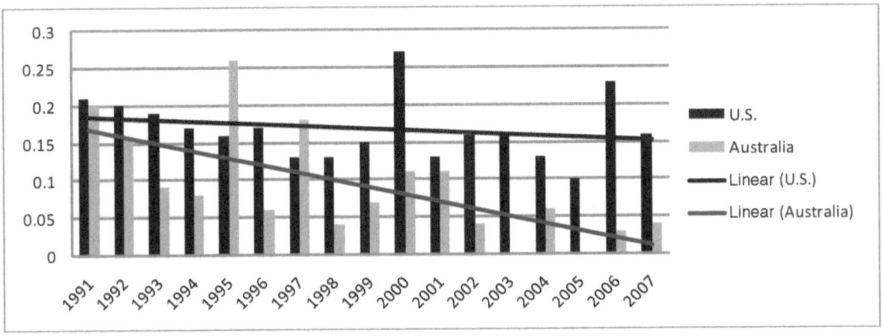

Figure 1.2 Coal mining fatal injury frequency rate comparison between the U.S. and Australia (1991–2007).
Source: Minerals Council of Australia, Safety and Health Performance Report, 1998–1999 and 2006–2007.

things changed during that period and these factors can hardly be controlled in research, it is almost impossible to approve the causal relationship between the change of regulatory approach and the improvement of safety performance in Australia. However, it is not difficult to see the correlation between the change of regulatory approach and the greater decline of fatality rate in Australia. Poplin et al. (2008) have a similar finding after analyzing annual lost-time injury (LTI) rates for coal mines in the U.S. with respect to Qld and NSW in Australia from 1996 to 2003. They find that LTI in all three regions was reduced. Qld and NSW, however, experienced a more remarkable decline (78% and 52%, respectively), compared with 20% in the U.S. They assert that the different regulatory approach "provides one explanation of the differential decline." (p. 1) Moreover, there is a general argument in scholarship that risk-based approach is more desirable than prescriptive regulation, although many problems in its implementation need to be fixed (see Chapter 2).

RESEARCH QUESTION

This research is not attempting to approve how risk-based governance plays as a causal factor in the improvement of mining safety—this would be an almost impossible job. Nor does it intend to add to the literature about the advantages and diadvantages of risk-based goverance, as well as how to make it work in practice, because existing studies have covered the topics very well. Rather, this book aims to explore the puzzle of why risk-based governance, which is very popular in many European and commonwealth countries, has been largely ingored in the U.S. Although the coal mining industry in the U.S. has introduced risk management to some extent, the progress is much slower than in many other countries. In Australia, risk management has been used extensively. Not only has it been voluntarily adopted by many coal mining companies, but it is also mandated by legislation in major coal mining states (Qld and NSW). In the U.S., despite some initial steps that have been taken by the federal regulator MSHA to move toward risk-based regulation between 2000 and 2005, these efforts largely failed, and the basic regulatory structure remains highly prescriptive. At the industry level, only a handful of large coal producers are in the process of developing risk management systems. This phenomenon is particularly surprising considering that risk management has been widely used, not only by the coal mining industries in many other developed countries, but also for many years by other industries in the U.S., such as nuclear and aerospace.

Therefore, the research question is why risk-based governance has diffused more slowly in American coal mining safety regulation. To answer this question, three sub-questions are examined. First, since this question implies that there is a certain degree of diffusion in the U.S., it is better to understand how and why the risk management model has been introduced to some extent in the U.S. coal mining industry before explaining why it moves slowly. Moreover, as previously defined, risk-based governance includes corporations' voluntarily adapted risk management systems and governments' risk-based regulation. Therefore, two additional sub-questions are: what are the barriers keeping the coal mining industry from fully embracing the risk management model, and what are the barriers keeping the U.S. government from moving toward risk-based regulation?

To better address the second and third sub-questions, the study adopts a comparative approach. The U.S. and Australia are compared at the industry and government regulatory levels. Moreover, to examine the second sub-question, a U.S. company that has introduced the risk management model is compared to another U.S. company that has not.

This research not only can help practitioners understand what prevents the diffusion of risk-based governance in the U.S. and how to move it forward, but it also has theoretical implications for recently growing literature on the diffusion of regulatory capitalism in general and risk-based governance in particular.

REGULATORY CAPITALISM AND RISK-BASED GOVERNANCE

Since the 1990s, people have increasingly recognized that liberalization and privatization have led not to deregulation but, ironically, to an expansion of state regulation. At the same time, there has been a concurrent expansion of regulation beyond the government undertaken by various non-state actors at the local, national, and international levels. Recently, some scholars have contended that such a new regulatory order is better conceived as "regulatory capitalism." Regulatory capitalism is different from "regulatory state" or "regulatory society," because it takes neither a state- nor a non-state-centered perspective (Braithwaite, 2008; Levi-Faur and Jordana, 2005).

A defining characteristic of regulatory capitalism is "reciprocal causation" among different components of the new regulatory order: government regulation of businesses as well as regulation within the state, corporations, and broader civil society (Braithwaite, 2008, 20; Levi-Faur, 2005, 20).[2] In particular, it emphasizes the role of government regulation as an initial power to generate change toward regulatory capitalism. For example, the growth of government regulation of businesses demanded more regulation inside the government to oversee government agencies (Hood et al., 1999). Moreover, as Braithwaite (2008, 16) discusses, government regulation enabled corporatization that "in turn enabled regulatory capitalism." On the one hand, the increase of government regulation has driven small firms that cannot meet regulatory requirements out of business. Added to the influence of antitrust and securitization, more and more mega-corporations have been created (Braithwaite, 2008).[3] On the other hand, because of greater concern with reputation, these mega-corporations, along with global or industry associations, have formed self-regulatory regimes to regulate firms in all facets of business, such as Responsible Care in the chemical industry (Braithwaite, 2008, 22). The increase of self-regulation in industry, then, enabled more involvement of nongovernmental organizations (NGOs) in providing standards and auditing, not only for businesses but also increasingly for government entities.

An embodiment of regulatory capitalism is the recent growth of risk-based governance, which is embedded in two developments. The first is the rise of systems-based management and regulation. As shown by the discussion of the reciprocal relationship between government regulation and restructuring in businesses, corporations have increasingly adopted internal regulatory systems to ensure their corporate social responsibility in the "shadow" of government regulation (Levi-Faur, 2005, 13).[4] As a response, government regulation becomes increasingly systems-based. Instead of requiring specific standards, government regulation encourages or mandates that businesses establish management systems to strengthen their internal regulatory capacities (Gunningham and Johnstone, 1999). The second development encouraging risk-based governance is an increasing focus on risk because of not only growing catastrophic risks in modern society (Beck, 1992), but also greater demands for transparency, accountability, and legitimacy of both government regulations and organizations (Hutter, 2005b; Power, 2007).

Hence, risk-based governance involves at least two interrelated components. On the one hand, many corporations have adopted internal risk management systems to identify, assess, and control risks in health and safety, environment, finance, and business operations in general. On the other hand, related to the idea of "systems-based regulation" (Gunningham and Johnstone, 1999), government regulations mandate or encourage the establishment of internal risk management systems. This book helps to better understand the diffusion of risk-based governance by examining the case of coal mining safety regulation.

DIFFUSION APPROACH

The emphasis on diffusion as an independent driver of the rise of regulatory capitalism is another defining characteristic of regulatory capitalism (Levi-Faur and Jordana, 2005). The role of diffusion has drawn the attention of regulatory scholars because the world is becoming increasingly interdependent, not only economically but also socially, with the advance of modern communication and transportation technologies. In such an interdependent world, decisions are made based on observations of others' actions in different countries and sectors. Therefore, Levi-Faur and Jordana (2005,8) note that to better understand regulatory change in the contemporary era, scholars need to go beyond structural approaches and focus more on "diffusion approaches, which embrace a combination of horizontal (country to country, or sector to sector), vertical (from global to local), and bottom-up (from the domestic to the international) explanations."

The study of the diffusion of regulatory capitalism is built upon literature on the diffusion of innovations and policy diffusion, including adjacent topics such as policy transfer, lesson drawing, and policy learning (Dolowitz and Marsh, 2000; Rose, 1991; Sabatier and Jenkins-Smith, 1988). While specific definitions vary in this literature, diffusion generally is considered as a process in which knowledge or information about ideas, technologies, policies, institutional arrangements, and organizational practices transfers among individuals or from one place to others.[5]

Moreover, such literature provides basic conceptual frameworks for diffusion studies. For example, Rogers's *Diffusion of Innovations* (2003) identifies five sequential stages in the innovation-decision process: knowledge, persuasion, decision, implementation, and confirmation.[6] More important to the research here, it points out the main elements that influence the diffusion rate of innovations. Some elements are related to characteristics of an innovation, and others are associated with the people who are involved in diffusion processes. For example, first, an innovation is more difficult to diffuse if it is perceived as being more complex and less compatible with existing culture and institutional arrangements. Second, communication channels, including mass media and interpersonal connections, are necessary to spread information about an innovation.

Although most studies of the diffusion of regulatory capitalism still follow the conceptual frameworks contributed by work on diffusion of innovations and policy diffusion, one influential study has gone beyond these basic frameworks and emphasizes detailed dynamics between individuals and structural changes. This work is Braithwaite and Drahos, *Global Business Regulation* (2000). Although the authors did not use the term at the time, this book examines how and why regulatory capitalism globalized. It also contributes a powerful framework, which allows scholars to explore dynamics that cross the micro and macro levels in the diffusion of regulatory capitalism. According to their framework, globalization of regulatory capitalism is "a process in which different types of actors use various mechanisms to push for or against principles" (p. 9).

A key of this framework is the theory of modeling. According to Braithwaite and Drahos (2000), modeling is the most important mechanism for the diffusion of regulatory capitalism. Because of its origins in psychology, particularly social cognitive theory that emphasizes reciprocal causation among environmental events, personal factors, and behavior (Bandura, 1986), the theory of modeling emphasizes the process of observational learning among interdependent people. Moreover, Braithwaite and Drahos (2000) go further and connect modeling with the politics of empowerment. The weak can be empowered by successfully creating an identity crisis and pushing a model as a remedy of the crisis (see Chapter 2 for more details about the Braithwaite and Drahos framework). The analytical framework of this study was developed based on the Braithwaite and Drahos framework.

SUMMARY OF FINDINGS AND CONTRIBUTIONS

Following the diffusion approach, particularly the theoretical framework introduced by Braithwaite and Drahos (2000), this research yielded several major findings. First, the risk management model has been to some extent introduced into the American coal mining industry because some of the largest coal mining companies, led by individual entrepreneurs—CEOs and senior safety professionals—have established commitments to higher safety standards. Their change of mentality has been both economically and socially driven. After this change, the logical next step is to search for new models. In this process, model mongers (safety professionals) and model missionaries and mercenaries (researchers and consultants) introduced the model of risk management into American coal mining companies and the industry.

Second, three major barriers keep the American coal mining industry from fully embracing the model of risk management. First, the existence of a large number of small operators, which lack resources and expertise to develop highly complicated risk management systems, prevents this model from being diffused in the industry. Second, increasingly prescriptive regulations have consumed the resources that companies could use to develop risk management systems and have created a mentality of compliance that is not compatible with the idea of risk management. Third, a group of model mongers, missionaries, and mercenaries have advocated a competing model—behavior-based safety—that is more attractive to the industry (see Chapter 2 for the definition of model mongers, missionaries, and mercenaries).

The third major finding is that the lack of three factors helps explain the failure of the U.S. government's move toward risk-based regulation: (1) strong imitative pressure from general OHS regulation; (2) strong model mongers, missionaries, and mercenaries; and (3) webs of dialogue.

This book makes both theoretical and practical contributions. From a theoretical perspective, it contributes to the scholarly literature about the growth of regulatory capitalism in general and about risk-based governance and the Braithwaite and Drahos framework in particular. Despite the increase of scholarship on the diffusion of regulatory capitalism in recent years, the current literature is still quite thin. The most important works are Braithwaite and Drahos, *Global Business Regulation* (2000), and 12 articles published in the *ANNALS of the American Academy of Political and Social Science* in 2005 where the term "regulatory capitalism" first was coined.[7] All of these articles, combined with Braithwaite and Drahos (2000), examined 13 sectors, including intellectual property, financial regulation, labor standards, environment, telecommunications, nuclear energy, food, and transport. Considering the scope and depth of the expansion of regulatory capitalism, these studies are far from sufficient to fully understand the rise of this new order. One important element that remains unexamined is the recent development of risk-based governance across countries and policy domains.

In the existing literature about risk-based governance, most studies focus on implementation, especially on the effectiveness of management systems. Surprisingly, few studies examine the emergence and development of risk-based governance itself. The handful of studies that do address such issues are weakened by taking a "bird's eye view" (Power, 2007, x); they fail to examine concrete mechanisms that lead to the growth of risk-based governance in particular settings. This book fills this gap by examining dynamics between individuals and structures in the development of risk-based governance in coal mining safety regulation. More important, the study also fills another significant gap in the current literature: the lack of understanding of the mechanisms that block the development of risk-based governance.

With regard to the Braithwaite and Drahos framework, this research focuses on a policy area they did not include. Moreover, application of this framework in existing literature is not extensive. In particular, although it has been more than ten years since the theory of modeling appeared in 1994, few scholars other than Braithwaite and Drahos have applied this theory to regulatory studies.[8]

In addition to these theoretical contributions, this book also may be of interest to practitioners in the U.S. coal mining industry. As a type of diffusion process, modeling involves knowledge transfer and interpretation. As such, modeling is both constituted by and constitutive of power relations.[9] Along with the increasing interdependence in the world, a country that more quickly adopts a model that is widely perceived as the best practice can gain a dominant position over laggards that do not participate in the discourse. In the era of information capitalism (Braithwaite and Drahos, 2000), countries compete not only for resources, economy, and technology, but also for modeling new ideas and practices.[10] In this case, because of the difficulty of embracing the model of risk-based governance, the U.S. has given away its leadership position in coal mining safety regulation to other countries such as the U.K. and Australia. These countries have produced extensive knowledge on risk-based governance that is followed by many other countries including the U.S. Therefore, this research can help policy-makers and practitioners in the American mining industry to understand why risk-based governance has moved slowly in the U.S. and what possible solutions might allow the U.S. to catch up in the race of modeling.

THE STRUCTURE OF THE BOOK

This book is organized as follows. Chapter 2 will give a more detailed review of the literature on risk-based governance, risk management, and the Braithwaite and Drahos framework. The purposes of the review are to clarify concepts, to define the standpoint and boundary of this research, and to explain the analytical framework adopted for field studies. Chapters 3 and 4 will answer the first and second sub-questions, respectively: how and why the risk management model has been introduced to some extent in the U.S. coal mining industry and what are the barriers keeping the coal mining industry from fully embracing the risk management model. Chapter 5 will explain why the Australian government successfully moved toward risk-based regulation. This chapter provides a comparative case that helps to answer the third sub-question about the barriers keeping the U.S. government from moving toward risk-based regulation, which is the subject of Chapter 6. Conclusions and possibilities for future research will be discussed in Chapter 7.

2 Risk-Based Governance and Analytical Framework

INTRODUCTION

As noted in Chapter 1, this book contributes to the literature addressing the rise of regulatory capitalism in general, and risk-based governance and the Braithwaite and Drahos framework in particular. Regulatory capitalism is like a river in which risk-based governance flows, while the Braithwaite and Drahos framework helps us understand how various streams converge into the river of regulatory capitalism.

Since Chapter 1 has discussed the characteristics of regulatory capitalism and the gap in literature on regulatory capitalism (the lack of understanding of the rise of regulatory capitalism, particularly some of its key components such as risk-based governance), this chapter elaborates further on literature on risk-based governance and the Braithwaite and Drahos framework. The purposes of this chapter are (1) to clarify key terms and definitions; (2) to define the standpoint and boundaries of this research; (3) to discuss further the gaps in existing literature; and (4) to develop the analytical framework for field studies.

RISK-BASED GOVERNANCE

Components of Risk-Based Governance

As previously discussed, risk-based governance involves interrelated components at the enterprise and government levels. At the enterprise level, many corporations have adopted internal risk management systems to identify, assess, and control risks for health and safety, environment, finance, and business operation in general. At the government level, there has been a growth of government internal risk management and risk-based regulation. Black (2005) termed these two strands "new public risk management." These strands are distinguished by the source of risk. For the former, risk arises from within governmental organizations, whereas for risk-based regulation, risk is seen as coming from the activities of regulated firms outside governmental organizations, which may prevent the government from achieving its objectives.

According to Black (2005, 516), risk-based regulation is a decision-making framework designed to prioritize regulatory activities and to deploy resources more efficiently, particularly concerning inspection and enforcement activities. Under such a framework, regulators are able to focus their limited regulatory resources on the greatest risks based on the assessment of the risk that government regulators would fail to achieve their objectives because of the failure or misconduct of regulated firms. Therefore, different from traditional regulatory approaches, the starting point of risk-based regulation is risks, but not rules (Black, 2010, 190). More specifically, risk-based regulation involves the following main elements. First, regulators determine what risks can be tolerated and at what level. Second, regulators rank risks of activities or firms based on risk assessment. Third, regulators decide how much enforcement resource needs to be deployed on certain activities or firms. In addition, to determine the best way to deploy regulatory resources, regulators also need to consider the allocation of risk responsibilities among governments, society, and even the regulated firms themselves. A way to allocate risk responsibility to regulated firms is to require them to establish internal management systems to identify, assess, and control risks. This is so-called "systems-based" (Gunningham and Johnstone, 1999) or "management-based" regulation (Coglianese and Lazer, 2003).

Therefore, it can be seen that the risk-based regulation and the risk management systems within enterprises, two components of risk-based governance, are interrelated. For example, under risk-based regulation, allocation of regulatory resources is determined by regulators upon the evaluation of the internal risk management systems in the firms. The better a firm's internal risk management, the lower the risk government regulators must address and the less inspection is needed. In addition, the two components of risk-based governance also overlap with the systems-based regulation previously mentioned. In the realm of occupational health and safety, for example, corporate internal risk management systems are the backbone of safety and health management systems that are required or encouraged by systems-based regulations, although safety and health management systems also include responsibility, communication, auditing, and training arrangements. Moreover, although risk-based regulation and systems-based regulation have different emphases (the former emphasizes developing and enforcing regulations based on risk assessment, whereas the latter stresses allocating risk responsibility to firms), these two approaches to a large degree overlap in practice.[1] Under systems-based regulation, regulators require businesses develop internal management systems for certain matters and the level of government intervention depends on how effective these management systems are. This process essentially involves an assessment of regulated firms' risks and accordingly allocation of regulatory resources. Similarly, under

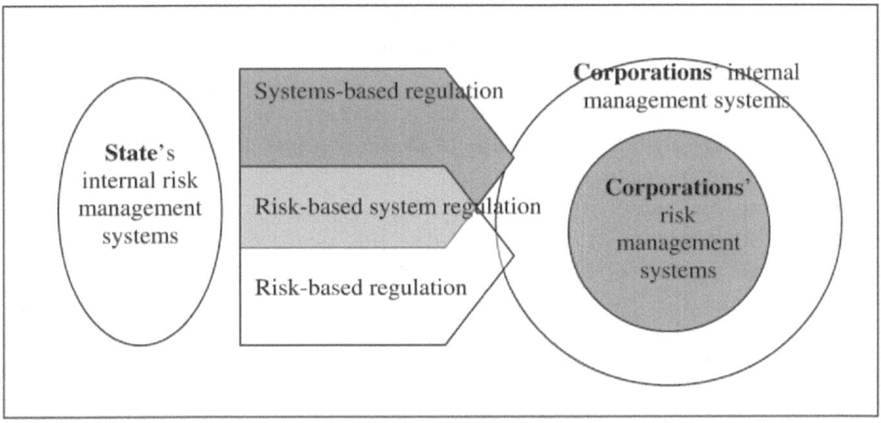

Figure 2.1 Interrelated components in risk-based governance.

risk-based regulation, a key indicator of evaluating regulated firms' risk is whether an effective management system is in place. The relationship of risk-based governance and systems-based regulation is shown in Figure 2.1.

This book focuses on corporate risk management systems and government regulations that mandate or encourage the establishment of internal risk management systems (components filled in gray color in Figure 2.1). I put these two components under the concept of risk-based governance. The government regulation that this research studies is closer to systems-based regulation than to risk-based regulation. However, the "system" of interest here is a risk management system, but not a broader health and safety management system. Hence, I use the term "risk-based system regulation" to refer to government regulations that require or encourage the establishment of corporate internal risk management systems.

This section gives an overview of the meanings of various interrelated concepts in current literature, including risk-based regulation, internal risk management systems, health and safety management systems, and systems-based regulation. This overview aims to help readers to better understand the standpoint and boundaries of this research in the voluminous literature regarding risk and regulation. Although risk-based governance has generated a large amount of research, this can be put into two major categories: the effectiveness of risk-based or system-based regulation and the development of risk-based ideas and practices itself.

Effectiveness of Risk-Based Governance

With the growth of risk-based governance and systems-based regulation in developed countries, many scholars seek to answer the question of whether these innovative approaches are actually effective. Although many report favorable results, the evidence is not sufficient to conclude that risk-based and systems-based approaches are effective in achieving their objectives. This is because it is extremely difficult to develop a methodologically rigorous study to separate the influence of the new approaches themselves from other contributing factors.

For example, according to Coglianese (2010), A management-based/systems-based approach to control risks has been introduced in the areas of industrial safety, food safety, and pollution prevention in the U.S. Since the 1980s, the Environmental Protection Agency (EPA) and the Occupational Safety and Health Administration (OSHA) in the U.S. require firms to implement "risk management planning" to assess and control chemical hazards. Firms are also required to audit their risk management systems on a periodic basis (p. 164). In terms of pollution prevention, a number of states have enacted regulations requiring firms using toxic chemicals to develop a management process to reduce polluting emissions. As for food safety, many countries, including the U.S., have adopted a systems-based regulatory regime called Hazards Analysis and Critical Control Points (HACCP), which requires firms to assess, monitor, and control potential hazards in all stages of food processing (p. 165).

Following the introduction of these systems-based regulations, the emissions of toxic chemicals, insurance claims in chemical industry, and food-borne illnesses have decreased significantly (Coglianese, 2010, 170). However, Coglianese claims that other factors may explain the declines to some extent. For instance, the reduction of food-borne illnesses may result from reducing pathogens in food handling or preparing, not just in food processing covered by the HACCP regulation (*ibid.*). Other changes in regulations may also contribute to the improvement. For example, the adoption of new performance-based hazardous air pollutant requirements by the federal government in the U.S. may cause a decline in toxic chemicals in the states that introduced systems-based regulations (p. 171). Moreover, Robson et al. (2007) conducted a systematic literature review of the effectiveness of occupational health and safety management system (OHSMS) interventions. They conclude that the existing studies are not sufficient to determine whether OHSMS is effective or not, although most of studies identified report positive results. This is because all studies examined except one have methodological limitations. The strongest research to date that supports the effectiveness of systems-based regulation is provided by Bennear (2007). Through sound statistical analysis, this research separates the effects of systems-based regulation from other factors, and compares states that adopt systems-based regulation in pollution prevention with states without systems-based regulation. The author found that the states with systems-based regulation have a greater trend of decrease in toxic emissions than others.

In addition to the literature that systematically evaluates the result of risk-based and systems-based approaches, many researchers put forward barriers in implementations that may prevent these approaches from achieving their intended objectives. The first is the lack of management commitment. Since many systems-based regulations simply require regulated entities to establish management systems, but not to implement them, employers who lack true commitment may produce a mere "paper system" (Coglianese, 2010). In studying environmental management systems (EMSs), Coglianese and Nash (2001) suggest a cautious view toward accepting EMS as a basis for public policy because the effectiveness of an EMS may depend on something other than the system itself, such as the management commitment to environmental improvements. In other words, a firm that has an EMS may not produce good environmental performance without true management commitment.

The second barrier is workforce ownership. Risk management is a comprehensive management system. Employees cannot understand and apply such a complex system in practice if they do not "own" the system. By examining the influence of enforced self-regulation on British Railway, Hutter (2001) concludes that the impact of risk management systems on everyday activities was limited although the systems were in place. This is partly because employees did not participate in safety practices actively and they had difficulty understanding the policies that were too voluminous and complex.

The third is the capacity of government regulators. The risk-based and systems-based approaches demand professionally qualified inspectors who can make judgments about whether a good risk management system is in place based on their expertise, not simply legal standards (Gunningham, 2006; Hopkins, 2007). These requirements for inspectors are totally different from those under a prescriptive approach. As Gunningham (1999) shows, the difficulty of enforcement is one of the reasons why trade unions tend to oppose the shift away from prescriptive regulation to a risk-based approach in Australian mining safety regulation. This is because, compared with complicated management systems, specific standards are much easier to enforce.

Despite the lack of strong evidence supporting the effectiveness of risk-based and systems-based approaches, and a series of problems that may prevent the achievement of their intended outcomes, there is a general agreement in regulatory scholars that risk-based and systems-based regulation are more desirable than traditional prescriptive regulation, particularly when the regulated sector is heterogeneous and social outputs are hard to define (Coglianese and Lazer, 2003; Coglinanese, 2010; Gunningham, 2004; Gunningham and Johnstone, 1999).

According to Gunningham and Johnstone (1999), regulatory standards can be classified into three major categories: specification, performance, and systems-based standards. Specification standards, which traditional prescriptive regulation relies on, tell employers precisely what measures and technologies to take, and allow little interpretation from employers. Such straightforward specification standards provide employees and inspectors greater opportunities to identify the requirements to employers and hold them accountable if standards are breached. However, this approach has a series of drawbacks. First, it tends to create a mass of detailed law. Second, it encourages a minimum compliance mentality and diminishes employers' responsibility for safety, because employers are required to carry out only mandated actions to fulfill their legal responsibilities. Third, since specification standards often encode engineering practices at the time that they were written, it eliminates operators' incentives to seek innovative and more cost-effective solutions. Fourth, it focuses largely on physical conditions that can be controlled through sound engineering principles, but fails to consider human and organizational factors.

In response to the limitations of specification standards, two alternative standards (performance and systems-based standards) have been increasingly attractive to scholars and practitioners. Unlike specification standards that tell employers exactly how to achieve compliance, performance standards specify desired outcomes or levels of performance. The virtue of this approach is that it can allow employers to choose particular ways to achieve the end, accommodate to changes in technology, and avoid the problems of "falling between the cracks" (Gunningham and Johnstone, 1999, 28). Nevertheless, compared with specification standards, performance standards may be hard to identify and enforce.

Despite their advantages, performance standards, as specification standards, may not encourage continuous improvement once certain outcomes have been achieved. In addition, they may not address businesses' internal culture and organizational arrangements. Hence, there has been a growing recognition of the importance of systems-based standards since the 1990s. The key of the systems-based standards requires or encourages employers to establish structures and procedures to assess and control risks, and to create an inbuilt system of maintenance and review. Similar to performance standards, systems-based standards allow firms the flexibility to choose their own means to health and safety. However, what distinguishes them from performance standards is their focus on businesses' internal organizational structures, responsibilities, procedures, and resources for implementing and maintaining an effective health and safety management system. Therefore, systems-based standards have the potential to stimulate modes of self-regulation and high-level commitment to safety within enterprises. They transfer primary responsibility of risk management to businesses. This allows people who work in the workplace to use their superior knowledge to develop plans to reduce risks. Moreover, employers are more likely to comply with rules and procedures developed by themselves than those imposed by government regulators. For government regulators, systems-based standards can reduce enforcement costs and tension between regulators and regulatees because regulated firms take more responsibilities for their own safety. In addition, the systems-based approach encourages the participation of nongovernment entities and workforce in regulatory processes; it is a more democratic approach than traditional top-down government regulation based on specification standards. According to Coglianese and Lazer (2003), the advantages of management-based/systems-based standards are particularly salient when the regulated sector is heterogeneous and social outputs are hard to define and measure. This is because the heterogeneity of the industry makes it impossible to develop universal technology (specification) standards that are applied to different firms over a long period. Similarly, the difficulty in defining and assessing social outputs makes performance standard impracticable. As far as the potential problems of risk-based and systems-based regulations discussed above, these can be dealt with through careful design and implementation. For example, to make sure the management systems adopted by firms are reliable, regulators can impose detailed record-keeping requirements and instituting third-party audits (*ibid.)*.

Besides the studies mentioned above that address the effectiveness of risk-based and systems-based approaches, another strand of literature examines the emergence and development of risk-based and systems-based ideas and practices.

Emergence and Development of Risk-Based Governance

Compared with the literature about the effectiveness of risk-based governance, fewer studies examine the emergence and development of this approach itself. Only a handful of such scholarly work has been identified (Frick et al., 2000; Hutter, 2005a, 2005b; Power, 2004, 2007).

With regards to health and safety management systems, *Systematic Occupational Health and Safety Management: Perspectives on an International Development* by Frick et al. (2000) gives the most comprehensive overview of the development, roots, and implementation of OHS management systems to date. According to these authors, the development of occupational health and safety management strategy follows at least three interrelated strands: (1) employers' voluntary introduction of highly formalized and documented occupational health and safety management systems; (2) mandatory requirements of OHS management systems in more and more countries; and (3) growth of standardization of safety management systems at both the international and national levels. This characterization reflects exactly the interrelated components of regulatory capitalism portrayed above: internal regulations within enterprises, governmental systems-based regulation, and regulations in civil society. Frick et al. (2000) also suggest that the following forces drive the development of OHS management systems. First, OHS management systems are extensions of quality control management into the field of OHS. Second, in the context of deregulation, OHS management systems are developed to justify reducing government regulatory scrutiny (pp. 4–5). Third, with increasing constraints on public finance, governments look for more cost-effective OHS strategies by placing more responsibilities on employers (Frick and Wren, 2000, 24).

In addition, two recent studies deal with the emergence and development of risk-based regulation and risk management. First, Hutter (2005a, 2005b) examines the emergence of risk-based regulation in financial services, environmental protection, and occupational health and safety across different countries. She puts forward two main clusters of explanations of the emergence of risk-based regulation. The first is a response to the so-called "regulatory crisis" of the 1980s and 1990s, when there were increasing demands from the public for legitimating regulations by introducing risk-based tools that were considered efficient, objective, and transparent. The second is related to changing economic and social conditions, particularly the rapid growth of new risks because of the evolution of scientific knowledge and technologies (Hutter, 2005b). Hutter's research did a good job of giving an overview of the logics behind the growth of risk-based governance. However, her explanations relied primarily on content analysis of documents and websites and did not show any concrete mechanisms and dynamics that led to the emergence of a risk-based approach. Having identified the necessity of additional research on this issue, Hutter (2005a, 14) suggests three areas for the research agenda. One item is systematic examination of the ways in which risk ideas have spread across countries and policy domains.[2]

The other study of note is Power's (2007) book *Organized Uncertainty*, the most systematic examination of the logic behind the development of risk management to this point. This book traces the distinct and related forms of logic and values that underlie the rise of the risk management idea. At the risk of oversimplifying Power's ideas, two types of logic lie behind the rise of risk management practices. The first is a functional explanation, which claims that the rise of risk management is a rational response to the fact that the environment has become more risky. This explanation also is expressed in terms of the logic of opportunity, claiming that risk management adds value and makes business sense. The second type of logic is a managerial logic of auditability and accountability (p. 23). This rhetoric claims that risk management emerges to meet the social expectations of being a good organization, in which risk must be seen to be managed. The logic of auditability is associated with the fear of blame by hostile stakeholders. Therefore, according to this point of view, the rise of risk management is less about risk and more about the management of legitimacy and reputation.

Although these works have traced the roots and logic behind the development of risk-based ideas and practices, they are weakened by taking a "bird's eye view" (Power, 2007), without examining the concrete mechanisms that lead to the growth of risk-based governance in a particular setting. For example, Power acknowledges that he owes "a deep intellectual debt" to institutional variety (2007, ix). He also notes that his book is not "a study of how one or more organizations adopt and adapt technologies . . . it is not about the process of negotiation between agents of change" (p. 24). This book seeks to fill this gap and address the questions that Power's book does not: the development of risk-based governance in a particular setting (coal mining safety regulation).

More important, this book fills another significant gap in the current literature—the lack of understanding of the mechanisms that block the development of risk-based governance. This neglect probably occurs because the rapid growth of risk-based governance across countries and domains has diverted people's attention from those cases where adoption of risk-based governance is difficult or fails. The only identified study relevant to this issue is Needleman's (2000) analysis of the institutional obstacles to OSHA's Cooperative Compliance Program (CCP) in the U.S. When proposing a future research agenda—systematic examination of the ways in which risk ideas and practices have spread across countries and domains—Hutter (2005a, 14) advocates that research "must take into account those domains and countries where risk ideas and approaches have not proved attractive and have not been adopted."

As discussed earlier, this book aims to examine concrete mechanisms that both drive and prevent the diffusion of risk-based governance in coal mining safety regulation, including corporate internal risk management systems. It is necessary at this point to clarify what risk management means in this study because the meaning varies across different countries, domains, and disciplines.

What Does Risk Management Mean?

The first issue that needs to be emphasized is that risk management involves multiple layers, ranging from major catastrophic risk management at the entire mine or plant site, to serious incident risk management for each project and change, to management of risks in routine tasks, and to an individual's risk management for each task performed at the face level (Figure 2.2). The top two layers emphasize the engineering design of physical systems, whereas the bottom two layers stress the risks related to workers' behavior when performing each task. The term "risk management" as used here refers to the whole structure of risk management, from the overall site level to the day-to-day face level. In practice, some use the term "risk management" mainly to indicate the bottom two layers of risk management, particularly when looking at workplace safety from a psychological or behavioral science point of view.

Moreover, risk management in this book refers not only to a holistic but also a formal, structured, and documented process. Based on the Australian Standard of Risk Management (AS 4360), Joy (2008) shows a typical process of risk management (Figure 2.3). The three steps in the box are the process of risk assessment, including risk identification, risk analysis, and risk control. Note that risk management involves not only risk assessment, but also a management system that includes the processes of communication, risk treatment, and monitor and review. In the risk treatment step, management needs to assign responsibilities to individuals to undertake control measures that management has decided to put in place.

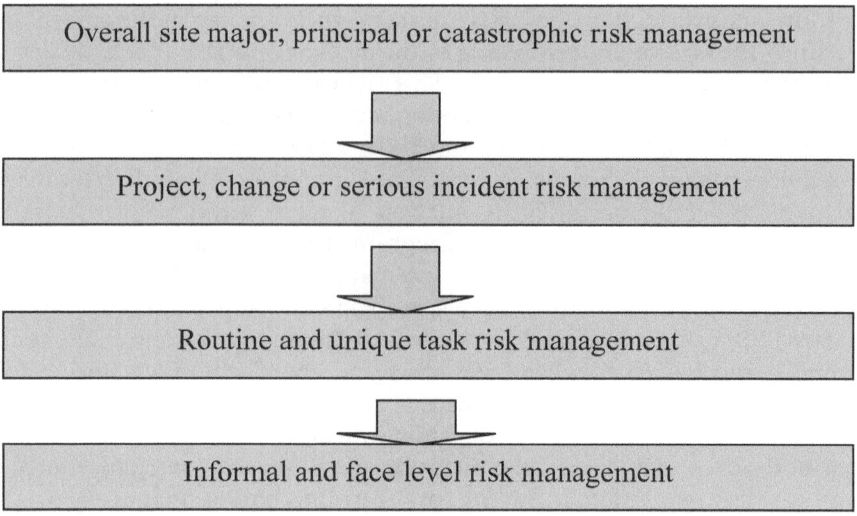

Figure 2.2 Multi-layered risk management.
Source: MSTTC (2006, 37).

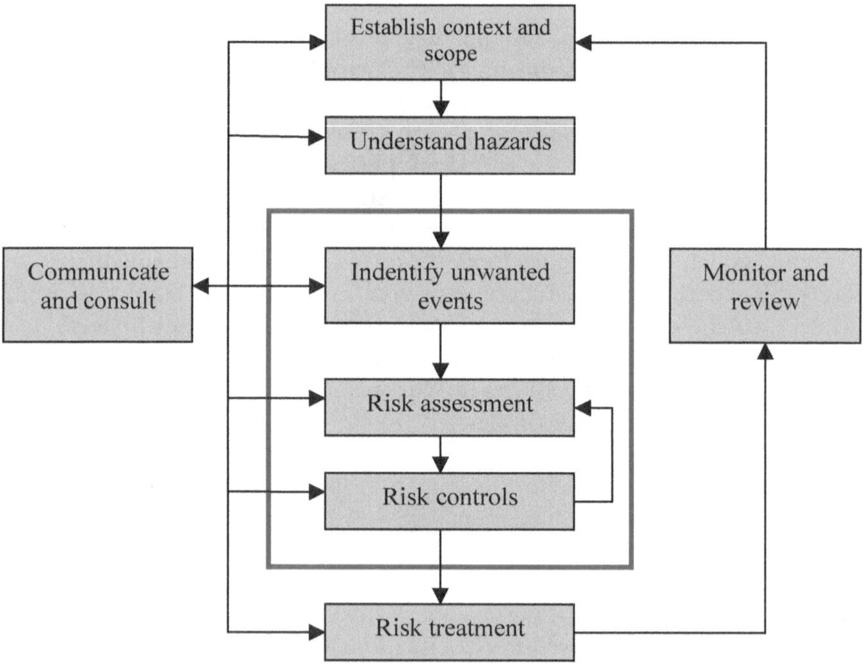

Figure 2.3 Risk management process.
Source: Joy (2008).

As just stated, risk assessment include three steps: risk identification, risk analysis, and risk control. *Risk identification* refers to identifying the hazards and the situations that have the potential to cause harm or losses (Joy, 2004). For example, a typical underground mine has 50–70 major mining hazards, such as explosions, fires, inundations, roof falls, and floods. A widely adopted method to identify potential hazards is to find unwanted release of energy. Sources of energy include gravity, electrical, mechanical, chemical, pressure, and thermal (Joy, 2004).

Risk analysis identifies the relative importance or priority of an undesired event. The priority is determined by the consequences and likelihood of each type of event. The relationship can be described by the equation:

$$\text{Risk} = \text{likelihood} \times \text{consequences}$$

Based on these two variables, a risk analysis matrix is widely used to rank risks. The two variables can be measured using qualitative or quantitative values. In qualitative risk analysis, the magnitude of likelihood and

Table 2.1 Qualitative Risk Analysis Matrix

Likelihood / Consequence	High	Medium	Low
High	1	1	2
Medium	1	2	3
Low	2	3	3

Source: Joy and Griffiths, 2007, 49.

Table 2.2 Hierarchy of Control

Eliminate Hazard	Eliminate hazards through change of equipment and mining process or method
Minimize or Substitute Hazard	When hazards cannot be eliminated completely, reduce or substitute hazards through engineering controls, e.g., improved ventilation, improved construction techniques, and improved equipment
Physical Barriers	Provide physical barriers that separate the hazard from the worker when energy has been released, e.g. roof rock reinforcement, sealing, refuge chambers, etc.
Warning Devices	Install device to monitor environmental/equipment conditions, e.g., gas monitors, alarms, and sirens
Procedures	Improve processes conducted by workers and management, i.e., inspections, audits, and trigger action response plans
Personnel Skills and Training	Provide training on skills, knowledge, behaviors, and observation of conditions

Source: Iannacchione, Varley, and Brady, 2008, 21.

consequences are classified by words, but not objective data. Table 2.1 is an example of the simplest form of qualitative risk analysis matrix.

Quantitative risk analysis can be applied when consequence and likelihood can be measured by numerical data. Quantitative risk analysis can provide more objective analysis because it relies less on people's judgment. Currently, most risk analysis in the coal mining industry is conducted qualitatively because of the lack of accurate quantitative data about event likelihood (Joy, 2004).

Having determined the importance of risks, the next step is to consider how to control them. *Risk control* follows a hierarchy of control strategy, which requires that risk analysis teams consider the measure at the top of the hierarchy first and leave the procedures and personal controls as the last choice (Table 2.2).

Moreover, a variety of techniques are commonly used in risk analysis. According to Joy and Griffiths (2007) and Iannacchione, Varley, and Brady (2008), several techniques, ranging from informal risk analysis to more formal and complicated techniques, include:

- Informal risk analysis: identifying and communicating hazards and risks in a task with no documentation.
- Job Safety Analysis (JSA): developing Standard Operating Procedures to identify and control potential hazards associated with each task.
- Workplace Risk Assessment and Control (WRAC): identifying and ranking potential hazards associated with each step of the mining process based on the project or geography of the mine.

Table 2.3 Workplace Risk Assessment and Control

Phase of Operation	Unwanted Event	Consequence	Likelihood	Risk Ranking

Source: Iannacchione, Varley, and Brady, 2008, 11.

Table 2.4 Preliminary Hazard Analysis

#	Unwanted Event	Likelihood	Consequence	Risk Ranking
1				
2				
3				

Source: Iannacchione, Varley, and Brady, 2008, 12.

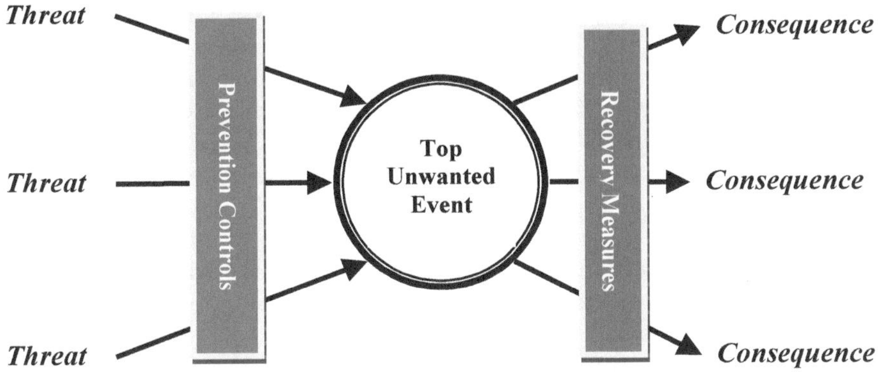

Figure 2.4 Bow tie analysis.
Source: Iannacchione, Varley, and Brady, 2008, 15.

- Preliminary Hazard Analysis (PHA): identifying and ranking potential hazards directly without breaking down the mining process.
- Bow Tie Analysis (BTA): based on a separation of the stages of prior and post-unwanted events, identifying threats (potential causes) that may lead to the hazard and possible control measures to prevent the occurrence of the unwanted event, and the consequences and possible control measures to reduce or minimize the loss, supposing an unwanted event has occurred.
- Fault Tree Analysis (FTA): starting from major unwanted events and tracing contributors to the unwanted events (usually used in quantitative methods).

In summary, this section has drawn the boundaries of this research—focusing on corporate risk management systems and risk-based system regulations. It has also reviewed and identified gaps in existing literature in these fields, and has clarified the meaning of risk management. The next section will further discuss the Braithwaite and Drahos framework, from which the analytical framework for field studies is developed.

THE BRAITHWAITE AND DRAHOS FRAMEWORK

As previously described, the rapid growth of risk-based governance across the world is clearly a product of diffusion. This book examines how risk-based governance diffuses in coal mining safety regulation in the U.S. and Australia, looking especially at the influence of interpersonal dynamics as well as the influence of these dynamics on institutional change. Braithwaite and Drahos (2000) provide a very powerful framework to examine such dynamics that cross the micro and macro levels. The framework is organized around three foundational concepts: actors, principles, and mechanisms. The authors argue that "regulatory globalization is a process in which different types of actors use various mechanisms to push for or against principles" (p. 9).

The *actors* in Braithwaite and Drahos's book include states, international organizations, individuals, NGOs, and epistemic communities.[3] From a diffusion perspective, two types of actors are most important: the individual entrepreneurs who initiate change and search for innovative solutions on the demand side, and experts who bring solutions to individual entrepreneurs on the supply side.

Principles in this framework are abstract ideas that guide conduct, which can unite a group of actors who share similar beliefs and languages. Braithwaite and Drahos (2000, 29) state: "the globalization of business regulation proceeds through contests of principles." They identify about 30 principles, including deregulation, national treatment, lowest-cost location, rule compliance, continuous improvement, and world's best practice. Principles are important because actors who desire certain goals may use principles instrumentally to reshape the regulatory world. More specifically, principles can be used to "ratchet-up" or "race-down" standards. When principles such as continuous improvement and world's best practice are proposed, standards will ratchet up. In contrast, rule compliance is often used to keep standards static. This echoes Powell and DiMaggio's (1991, 248) emphasis on the importance of institutional logic, defined as the "organizing principle which is available to organizations and individuals to elaborate." Conflict occurs when contradictory institutional logics exist. Individuals or organizations can advance their interests by defining and elaborating different institutional logics.

As previously stated, actors may drive up standards to obtain their desired goals. These goals can be economic, such as savings on waste, early-mover advantages, and other competitive advantages; they also can involve social considerations, such as higher reputation and legitimacy (Braithwaite and Drahos, 2000,560). Braithwaite and Drahos note:

> Where regulated communities have realized that the fate of one can determine the fate of all, the principle of deregulation has dropped out of the regulatory discourse altogether. . . . The standard of safety will go up rather than down . . . although the world's best practice principle is a costly principle for actors to follow, it is also an asset-building principle. It helps to create the asset of reputation. Where the best practice principle is weak, it is because the driver of reputation matters less and immediate costs determine all. (pp. 515–519)

As Braithwaite and Drahos (2000) state, actors use not only principles but also a series of *mechanisms* to achieve desired goals. These mechanisms include military coercion, economic coercion, systems of reward, reciprocal adjustment, non-reciprocal coordination, capacity-building, and modeling. Although globalization of regulation is not a result of a single mechanism, Braithwaite and Drahos (2000) contend that modeling is the most important mechanism, and it can be found in all of the cases they studied. Modeling is defined as "actions that constitute a process of displaying, symbolically interpreting and copying conceptions of action" (p. 581). Therefore, modeling includes the process of both model supply and adoption. In addition, modeling is not simply an imitation, but also involves cognitive processes.

Braithwaite and Drahos's (2000) theory of modeling has its origins in psychology, particularly social cognitive theory that emphasizes reciprocal causation among environmental events, personal factors, and behavior (Bandura, 1986). According to Bandura, people learn new knowledge, skills, and behavior mainly through observational learning. One special type of observational learning is enactive learning. In other words, people learn from observing the effects of their own actions. If enactive learning that involves repeated trial-and-error experience was the only way that people acquired knowledge, learning costs would be too high, and social development would be very slow. In fact, on most occasions, people learn through modeling (the most important type of observational learning). Learning through modeling is not only about observing others' performance and consequences, but also involves transforming observed information into symbolic form that can be used as a guide for future actions and can be transmitted to wider society beyond observers' "immediate environment" (Bandura, 1986, 47). Therefore, the theory of modeling is able to reveal the interpersonal dynamics of diffusion and the influence of those dynamics on institutional change.

According to Bandura (1986), the study of observational learning needs to account for both the success and failure of modeling. Based on diffusion theory, observers have to go through two stages to successfully adopt a new model: acquisition of knowledge and adoption of the model in practice (p. 144). A model may fail to be adopted simply because observers are not exposed to the model or because they lack the cognitive skills to transform it into symbolic representation. Moreover, even though observers have acquired knowledge, they may not turn knowledge of the model into actions. In both stages, there are many impediments of modeling, including the attributes of the observers and the properties of the model itself. For example, observers may lack social networks to access the information; their cognitive competencies may not allow them to perceive the benefits of the model; and their cognitive structures may lead them to look for some things but not others. In addition, a model that has the property of generating immediate and observable rewards is more likely to be adopted in practice. The theory of modeling has the capacity to explain not only drivers of, but also barriers to diffusion. Therefore, it fits very well with the purpose of this research—exploring the factors that prevent the diffusion of risk-based governance in American coal mining safety regulation.

Braithwaite and Drahos (2000) go further and connect modeling with a politics of empowerment that involves several types of actors: model missionaries, mercenaries, mongers, misers, and modernizers. Model *missionaries* are people who spread the model around the world, not only because of monetary incentives, but also because of belief. Model *mercenaries* are promoters of models for commercial purposes. According to Braithwaite and Drahos, the work of model missionaries and model mercenaries can only start the modeling process. The most important step is to turn modeling into political processes through model mongering. As per Braithwaite and Drahos (2000, 585), model *mongers* are agents who pursue their political agenda by adopting multiple models. They will devote more resources to develop the model that attracts political support. The model developed by model mongers can gain a dominant position if it is accepted by model misers and model modernizers. Model *misers* are adopters who realize that the adoption of an existing model is more efficient than reinventing the wheel. Model *modernizers* are those who adopt models for identitive concerns. They want to be perceived as "modern" and advanced. Note that in practice an individual may have the characteristics of multiple types of actors. For example, scholars are increasingly involved in consulting services. They may be both model missionaries and mercenaries.

The Braithwaite and Drahos theory of modeling, particularly its connection with the politics of empowerment, is very powerful in explaining model diffusion at the level of government regulatory policy. However, the authors also stress that modeling can occur on a firm-to-firm basis (p. 542). In this case, model mongers particularly refer to firms' internal entrepreneurs, who pursue political agendas within companies.

Moreover, modeling is both constituted and constitutive. The power of a model is that it can set the terms of debate when it is put successfully on the table. On many occasions, when individuals get into a regulatory game, they have to face an institution embedded in a model adopted earlier. It is against the existing model that people define their identities and interests. Braithwaite and Drahos (2000, 581–582) argue that:

> More of the variance in culture and institutional structure is explained by patterns of modeling rather than by configurations of actor interests and their relative power. The clash of interests is not only quantitatively less important as an explanation, it is causally secondary: the interests of the actors are constituted in response to received models; they do not pre-exist.

Integrating the concepts of actor, principle, and modeling, Braithwaite and Drahos (2000, 561–562) present two different explanatory sequences of regulatory change: proactive and reactive. In the proactive sequence, change starts from individual entrepreneurs who propose regulatory innovation. They enroll organizational power and model regulatory innovation.

In order to achieve competitive advantage, early movers push to globalize the innovative regulatory standards they have embraced. In the reactive sequence, they contend that strategic individual entrepreneurship from businesses or NGOs is less important. More important are key individual entrepreneurs who have direct organizational control of the state's capacity. They act in response to disasters that mobilize the mass public that demands change. The proactive sequence better explains the bottom-up process of regulatory change starting from enterprises or industry, whereas the reactive sequence better fits the top-down process of regulatory change initiated directly by governments.

As previously noted, an advantage of the diffusion approach is that it allows researchers to look beyond structural forces of change. The diffusion approach, however, does not deny the influence of structures. There are limitations to simply using either micro or macro explanation alone. For example, it is inadequate to see structural variables as having determinative influence and to ignore the social and political constitutive processes of individuals. It is equally inadequate to forget that individuals' actions in the modeling process are facilitated and constrained by structural variables. Hence, Braithwaite and Drahos (2000, 582) advocate viewing modeling and structural explanations as complementary.

An important structural variable that Braithwaite and Drahos (2000) suggest in the process of regulatory globalization is the "web of dialogue." According to them, the web of dialogue is more important than the web of rewards and coercion for three reasons. First, unlike the web of rewards and coercion that can be used only by the strong, the web of dialogue is suitable to both the strong and the weak. Second, the web of dialogue can help to identify problems, build normative commitment, reach consensus on principles and norms, and institutionalize praise and shame. These benefits can hardly be achieved through the web of rewards and coercion. Third, and more important to this research, the web of dialogue can facilitate modeling (Braithwaite and Drahos, 2000, 32).

Most webs of dialogue identified in the Braithwaite and Drahos case studies occur at the international level. Therefore, Braithwaite and Drahos do not pay adequate attention to how other structural variables influence the formation of webs of dialogue, such as national culture, political structures, and industrial relations. The influence of such structural variables on the formation of webs of dialogue, however, cannot be ignored for a study like this research that focuses on situations within a country.

Some scholars have pointed out that a systems-based approach is likely to succeed in a regulatory culture that relies more on a cooperative approach to policy formation and implementation and is less likely to be effective in environments characterized by conflict, distrust, and a deterrent style regulation (Gunningham and Johnstone, 2000; Needleman, 2000). As many analysts suggest, American regulatory culture is characterized by a distinctive "adversarial regulation" or "adversarial legalism" as opposed to more "collaborative regulations" in other industrial countries (Gunningham and Johnstone, 2000; Kagan, 1991; Kelman, 1981; Vogel, 1986). As such, I expected that the adversarial regulatory culture in the U.S. would have a strong influence on the formation of the web of dialogue, and therefore would be a factor that impedes the diffusion of risk-based governance.

According to Kelman (1981, 162), in a country with a collaborative regulatory style, the government tends to rely on a small group with a limited number of representatives of major stakeholders to draft laws. In a country with a smaller population, these representatives often know one another personally. This type of small group with representatives who meet privately and informally is a useful forum for closed-door negotiation and consensus-building (Kagan, 1991, 373; Kelman, 1981, 132). Moreover, the representatives tend to be more independent from their stakeholder principals who delegate their authority when they work in a small group. Hence, they can to a greater degree make decisions based on their own beliefs, not just what those they represent want (Kelman, 1981, 151). In addition, government inspectors tend to use more supportive and persuasive strategies to enforce laws. Inspectors usually are assigned to a specific set of workplaces for a long time; therefore, they can gradually develop close working relationships with businesses (Kelman, 1981, 182).

By contrast, in the U.S., dominated by adversarial regulation, both bill preparation and administrative rulemaking rely on a "judicialized" style of public hearings, public notice and comments, and formal responses to interest group arguments (Kagan, 1991, 373–374). Kelman (1981, 169) contends that this is because government authority is viewed as being legitimate if it appears as a neutral judge. Under such a judicialized rulemaking process, each party presents its opinions and evidence with little dialogue among the various groups.[4] Moreover, in an adversarial regulatory system, the primary enforcement strategy is deterrence based on suspicion of businesses' willingness to comply with the law. The major purpose of inspection is to search for violations. Government develops highly specific rules, and inspectors simply "go by the book" and cite all violations observed (Gunningham and Johnstone, 2000, 69). The law tends to be highly specific because there is wide suspicion that inspectors may run out of control and be captured by the regulated if more discretion is granted (Kelman, 1981, 191). Under such an adversarial enforcement philosophy, it is difficult for government inspectors and businesses to cooperate. Moreover, any shift toward a collaborative approach is politically dangerous.

Adversarial regulation in the U.S. stems not only from the country's self-assertive value and tradition of individualism, but also from fragmented governmental authority (Kagan, 1991; Kelman, 1981). According to Kagan (1991, 370), American government is designed based on deep suspicion of central government authority that may cause tyranny. It strongly emphasizes dispersed government powers, checks and balances among government bodies, and citizens' rights to challenge governmental decisions in court. No government body in the U.S. has sufficient discretionary authority to enforce a compromise among contending political interests (*ibid.*). Moreover, a particular interest group can access various government bodies other than the relevant executive agency to pursue its interests, including Congress and the courts. Because of the separate elections of members of Congress and the president, members of Congress are independent political forces who can influence the direction of regulation and agency behavior (*ibid.*).

A parliamentary political system, on the contrary, often features a strong and highly professional central bureaucratic authority. Law grants regulatory agencies broad discretion so that they can have the final say (Kagan 1991, 398). In addition, the influence of parliament is relatively small since the party in office controls the majority of seats in parliament (Kelman, 1981, 16). Legislation that is prepared by executive agencies is often automatically approved by the parliament, and courts rarely reverse government decisions. As Kagan (1991, 398) states, in collaborative regulatory systems, participants in small working groups know there are no allies in the court or legislature where they can turn for help. They also know that the agency will decide if no agreement is reached. This situation compels them to sit down at the table to negotiate seriously. Hence, this type of centralized government authority tends to foster the formation of webs of dialogue that more fragmented government structures tend to undermine.

Without dismissing its advantages, adversarial regulation may cause a series of problems, such as costly and time-consuming policy implementation and legal deadlock (Kagan, 1991, 370–377). However, the most important disadvantage is its "corrosive effect on personal and institutional relationships" (Kagan 1991, 378). In such a regulatory culture, different parties will not cooperate wholeheartedly and exchange information. Instead of seeking to align their desires with broader public interests, they tend to act opportunistically by distorting information and using legal action to delay policy implementation in order to gain advantage over the other side (Kagan, 1991, 389).

Although the Braithwaite and Drahos framework is very powerful, it has not generated wide applications in in-depth case studies. This is probably because the application of this framework requires very detailed and micro-level information that is hard to collect. In particular, I have identified no application of the theory of modeling in regulatory studies by scholars other than Braithwaite and Drahos in the more than ten years after it was first published in 1994. Scanning the Web of Science: ISI Citation Databases yielded only 13 articles that cited Braithwaite's (1994) article. Among these, only two published by Drahos were related to regulatory policy. This is clearly a significant gap that needs to be filled.

This section has discussed the Braithwaite and Drahos framework, the gap in existing literature related to this framework, as well as the literature on adversarial regulation. Based on this literature, I developed the analytical framework for fieldwork designed to explain the growth of risk-based governance and, more important, the barriers to its diffusion in American coal mining safety regulation.

ANALYTICAL FRAMEWORK

Here, risk-based governance refers to corporate internal risk management systems and government risk-based system regulations that mandate or encourage the establishment of such internal systems. As part of the order of regulatory capitalism, there is no doubt that mutual influence exists between government regulation and enterprises or the industry's voluntary adoption of risk management. However, governments and enterprises and the industry often have different agendas in determining if a new practice will be adopted or not. Therefore, for analytical purposes, this research separately examined the development of risk-based governance at the enterprise and industry level as well as the government regulatory level.

At the enterprise and industry levels, the process of adopting risk management starts with individual entrepreneurs, such as chief executive officers (CEOs) and safety professionals, who advocate a proactive culture of accident prevention. These entrepreneurs realize that ratcheting-up safety standards in the industry is needed for both economic and identitive purposes. They enroll organization power and uphold principles such as continuous improvement and sustainable development. Having established the commitment to higher standards, they search for new models to implement that commitment. Model missionaries and mercenaries in epistemic communities come in to change the values in the industry and to spread knowledge about new models. Model mongers—safety professionals within companies—try to put the change on the companies' agenda and to develop new safety standards within firms based on the new model adopted. Then, those early movers push to spread the new model to the whole industry and to incorporate their standards into government regulation. Hence, modeling is part of the process of driving up standards. In addition, the modeling process is influenced by structural variables such as industry structure and regulatory climate (Figure 2.5).

At the government regulatory level, governments see the need to change after a disaster occurs. When this "policy window" opens (Kingdon, 1995), model mongers, particularly policy entrepreneurs in the government, will suggest existing regulatory innovations that have been adopted in other settings. If a model is accepted by the model misers and modernizers, the change of regulations will occur. However, when the policy window opens, there may be counteracting forces driven by people who support the status quo by advocating oppositional principles. They also try to put their preferred regulatory model on the agenda. Whether the new model mongers can gain the dominant position depends on whether they can reach agreement with contending groups, whether model modernizers or misers

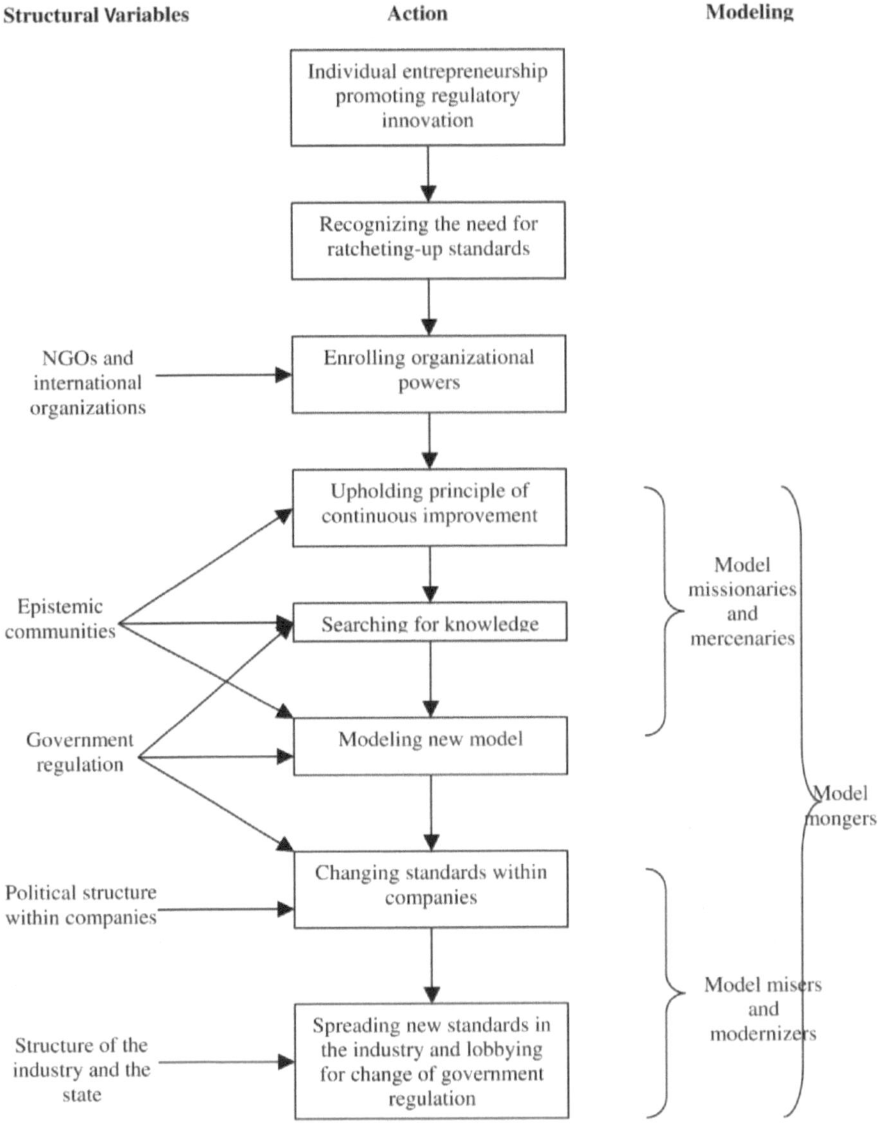

Figure 2.5 Process of driving up standards and modeling at the enterprise and industry levels.

support the new regulatory model, and whether model missionaries and mercenaries in the epistemic community can help the different parties to achieve shared understandings. Therefore, a structural variable is important to the model mongering process: the web of dialogue. The formation of the web of dialogue is, in turn, influenced by other structural variables such as industrial relations, an adversarial or cooperative culture, and the nature of the existing political system (Figure 2.6).[5]

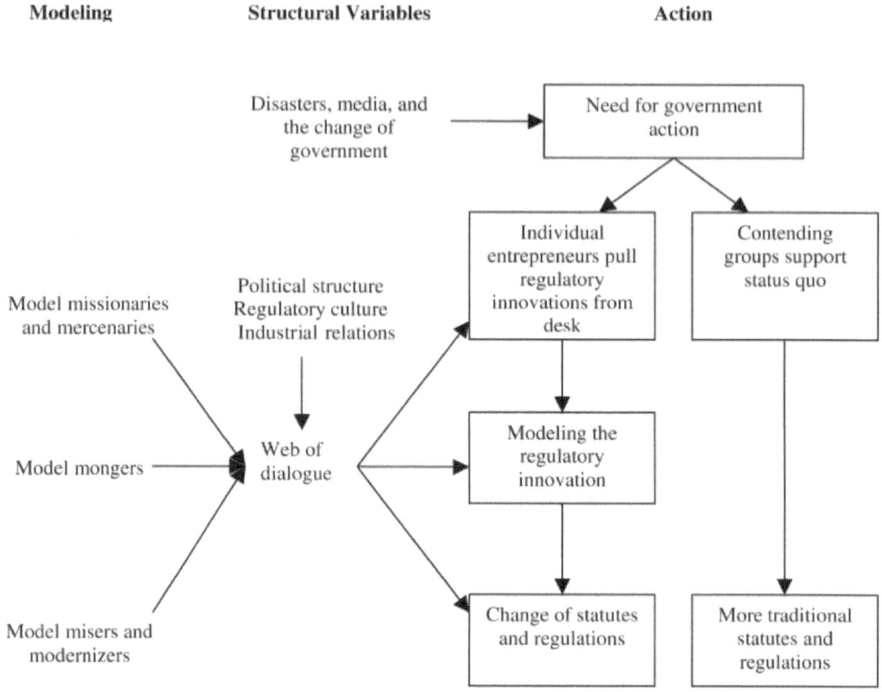

Figure 2.6 Process of driving up standards and modeling at the government policy level.

CONCLUSION

This chapter has reviewed the literature that this study builds upon and seeks to contribute to: works addressing the growth of risk-based governance as well as the Braithwaite and Drahos framework for studying regulatory globalization. It has also defined the boundaries of the research, focusing on risk-based governance, including corporate internal risk management systems and risk-based system regulation. I also elaborated on the significance of this research, clarified the meaning of risk management, and described the analytical framework that guided the field studies. Based on this framework, the following chapters will get into the substantive analysis of the findings in studying the three sub-questions of this research.

3 Diffusion of the Risk Management Model in the U.S. Coal Mining Industry

INTRODUCTION

The main question that this book seeks to answer is why risk-based governance evidently has diffused more slowly in U.S. coal mining safety regulation than in many other developed countries. This question implies that there has been a certain degree of diffusion of risk-based governance in the U.S. The purpose of this chapter is to answer the first sub-question of the research: how and why has the risk management model been introduced in the coal mining industry in the U.S.? This chapter also illustrates concrete mechanisms that led to the growth of regulatory capitalism.

Following the analytical framework developed in Chapter 2, I find that the diffusion of risk management in the American coal mining industry is part of the collective process of driving up standards. The process started from the change of mentality in some of the largest coal mining companies. Driven by both economic and social considerations, particularly the emerging recognition of common fate, and led by individual entrepreneurs—CEOs and senior safety professionals—companies have established a commitment to higher safety standards. After the change of mentality, the next logical step is to search for new models to fulfill the commitment. In this process, model mongers (safety professionals) and model missionaries and mercenaries (researchers and consultants) introduced the model of risk management into American coal mining companies and the industry as a whole.

This process of driving up standards in the American coal mining industry is not unique. Similar processes have occurred in many other industries and in the mining industry at the global level. In this chapter, the process of driving up standards collectively in the mining industry at the global level will first be briefly described to set the scene. Then, the process of driving up standards and modeling risk management in the American coal mining industry will be discussed. For analytical purposes, modeling risk management at the enterprise and the industry levels will be examined separately. The influence on individual companies of modeling risk management at the industry level then will be addressed.

DRIVING UP STANDARDS IN THE MINING INDUSTRY AT THE GLOBAL LEVEL—A MOVE TOWARD SUSTAINABLE DEVELOPMENT

As Braithwaite and Drahos (2000) suggest, regulatory globalization is a process in which different types of actors use various mechanisms to push for or against principles in order to obtain desired goals. The desired goals can be both economic and normative. Braithwaite and Drahos also indicate that when a common fate among all the actors is recognized, regulated communities tend to advocate principles to build an asset of reputation and collectively drive up standards. Such a process has taken place in various industries.

The best example is the establishment of the Institute of Nuclear Power Operations (INPO), an industry self-regulatory organization in the U.S. nuclear industry. According to Rees (1994, 43–45), the creation of INPO is a response to two "fundamental threats" to the survival of the whole industry after the Three Mile Island (TMI) event: additional government regulation and unsafe nuclear utilities. After TMI, the Nuclear Regulatory Commission "prepared to launch a massive regulatory assault on the industry" (p. 43). In particular, many regulatory initiatives targeted "institutional arrangements for managing, operating and maintaining" nuclear plants (p. 22). In order to prevent further government intervention into management, INPO was created to voluntarily embrace "industry-wide self-regulation" (p. 44). This process reflects the constitutive role of government regulation for regulatory capitalism discussed previously. Moreover, the industry realized "common interests and mutual interdependence" among all utilities because a catastrophic accident would lead to public opposition to nuclear power as a whole (p. 43). In other words, the failure in one operation can "destroy the credibility of all the others" (p. 44). All the actors in the industry became "hostages of each other" (p. 45). So INPO was created to foster a community in which all utilities that previously ran independently came together to work collectively to achieve "a standard of excellence" as opposed to a minimum compliance mentality (p. 41).

The U.S. coal mining industry obviously has not reached the stage of nuclear industry described above, where strong common interests bond all utilities together. This is partly because companies in the mining industry are much more diverse than those in the nuclear industry, ranging from mining companies with thousands of employees to small mines with only a dozen workers. Since the 1990s, the mining industry, however, has increasingly realized some common threats facing the whole industry that have damaged the reputation of the sector and have undermined the social license to operate.

The coal mining industry today operates in a changing environment, in which there are growing stakeholder expectations of businesses' environmental and social performance.[1] For instance, NGOs are more active, focusing on businesses' social responsibilities. The financial service industry has increasingly linked investment to businesses' social and environmental records. Employees are more interested in safe working environments. The influence of these stakeholders has been aided by changes in the environment in which mining companies operate. The first is increased public scrutiny of mining activities because of the development of new communication technologies (MMSD, 2002). Mining companies today operate in a world where there is no place to hide. Second, there is increasing competition among mining companies to access mineral resources because of "competing values over the best use of land" and stringent requirements by stakeholders and governments (Walker and Howard, 2002, 12). Therefore, as a result of these pressures, the mining industry has recognized the need to improve its reputation and to maintain social license to operate.

The solution that the mining industry has adopted to improve its reputation is to embrace the principle of sustainable development, which requires firms to go beyond corporate responsibility purely for the profit of shareholders and to consider the interests of a broader range of stakeholders who may not have direct economic ties with the firm, including employees, consumers, communities, and environmental groups.

Consistent with the analytical framework, the mining industry's move toward sustainable development was led by individual entrepreneurs—the CEOs who had recognized the need for driving up standards in response to the fundamental threats to the industry. They created and enrolled organizations to advocate the principle of sustainable development and subsequently to develop rules to implement this principle.

According to a series of reports of the Mining, Minerals and Sustainable Development (MMSD) project, the major shift toward sustainable development started in 1999 in Davos, Switzerland. At the 1999 World Economic Summit, nine chief executive officers of the world's largest mining companies came together and discussed the challenges to the industry's "social license to operate." Cooperating with the World Business Council for Sustainable Development (WBCSD), they initiated the Global Mining Initiative (GMI). According to Sheehy and Dickie (2002, 15), GMI had three major tasks: (1) reviewing international minerals sector associations; (2) undertaking a major independent study of the sustainable development challenges confronting the industry; and (3) convening a major industry conference in Toronto in May 2002.[2]

In order to keep the study independent from the mining industry's influence, the leading companies decided to transfer the second element of GMI to a two-year independent research project—the MMSD project. In 2000, again through WBCSD, they contacted the International Institute for Environment and Development (IIED), an independent international research organization in sustainable development, and asked it to manage this study. This project sought to assess the current situation of the industry in transition to sustainable development, to provide an action plan for how the mining and mineral industry could best contribute to the global transition to sustainable development, and to build a platform for ongoing dialogue among stakeholders (MMSD, 2002).

The MMSD project produced a final report—*Breaking New Ground*. It proposed a four-step action plan following a sequence of developing principles, creating rules for implementing principles, and building capacity to implement rules. The report emphasized that actions should be taken collaboratively among those having common interests through existing or newly created associations or networks. The report recommended the International Council on Mining & Metals (ICMM) take a leading role in developing a Declaration on Sustainable Development. Companies were encouraged to sign the Declaration. In order to implement this commitment, the report suggested developing a Protocol including a specific and measurable system of third-party verification.

The later development followed exactly what was recommended in the MMSD report. ICMM member companies signed the Declaration on Sustainable Development at the Global Mining Initiative conference in Toronto in May 2002. After the Toronto Declaration, ICMM gradually developed the protocol mentioned above, which is called the ICMM Sustainable Development Framework. It includes three parts: ten principles, reporting in accordance with the Global Reporting Imitative framework, and independent assurance. All ICMM members have committed to this mandatory framework.

Among the ten principles, two are directly related to this research:

> Principle 4: implement risk management strategies based on valid data and sound science;
>
> Principle 5: seek continual improvement of our health and safety performance. (ICMM, 2009)

ICMM is also a CEO-led industry organization formed in response to threats to the industry. It was established to represent 16 of the world's leading metal mining companies and associations in 2001. Currently, ICMM members include 19 of the largest companies and 30 associations. According to the information on its website, the mission of ICMM is twofold. The first is to make a contribution to raising standards across the industry as a whole. The second is to distinguish its members as industry leaders. This is because these ICMM member companies realized that the efforts of industry leaders were unrecognized by the public, and the image of the industry was, to a large extent, shaped by laggards (Walker and Howard, 2002, 11), because of the lack of a differentiation mechanism to separate good from bad performers. Today, on the websites of these ICMM member companies, there is one link entitled sustainable development. The annual sustainable development report is available on the website, which contains the performance on major issues such as health and safety, environmental protection, and community development.

The growth of sustainable development at the global level has led to a dramatic dissemination of risk management across countries. This is because risk management is considered a key tool to achieve the principle of sustainable development. As previously stated, one of the ten principles in the ICMM sustainable development framework is to "implement risk management strategies based on valid data and sound science" (ICMM, 2009). ICMM is one of the key promoters of the knowledge and best practices of risk management. A major initiative is ICMM's sponsorship of the Minerals Industry Safety and Health Centre (MISHC) at the University of Queensland in Australia, which began in 2005. The aim of this sponsorship is to add international data about health and safety risk management into the Minerals Industry Risk Management Gateway (MIRMgate) data system.[3]

The mining industry's move toward sustainable development is a typical example of the reciprocal causation among different components of regulatory capitalism. In response to the pressure imposed by government regulations and civil society, large mining companies, led by individual entrepreneurs who recognized the legitimacy crisis that the industry encountered, collectively drove up standards by upholding the principle of sustainable development. They enrolled international and research entities in order to promote new regulatory strategies, including risk management. The recent growth of risk management in the American coal mining industry follows a similar logic and process.

MODELING RISK MANAGEMENT IN THE U.S. COAL MINING INDUSTRY

Modeling risk management has occurred both in individual companies and in the industry as a whole. These two tracks of development interact with each other. For analytical purposes, the modeling in individual companies and at the industry level will be discussed separately in the first two parts of this section. In each part, the modeling processes will be examined following the discussion of a change of mentality. This is because searching for new models is the logical result of the change of mentality committing to higher safety standards. As Parker (2002, 31) states, to create an "open corporation" that can respond to social responsibilities, corporate management has to go through three phases: the commitment to respond, the acquisition of specialized skills and knowledge, and the institutionalization of purpose. This two-step process—commitment and search for models—can be applied not only to an individual company, but also to an industry as a whole. The third part of this section addresses the influence of modeling at the industry level on individual companies.

Modeling Risk Management in Individual Companies

Change of Mentality

Unlike the mining industry at the global level, the U.S. coal mining industry has not systematically incorporated workplace health and safety and other social and environmental issues into an overall principle of sustainable development. However, there has been a growing commitment to improved workplace health and safety, at least within some of the largest mining companies. Recently, some companies such as Consol Energy have built up the commitment that only zero accidents are acceptable.

As a respondent stated in his remarks at the Bluefield Coal Symposium in 2008, the major difference of the industry today is "culture and expectations." In a personal interview he further explained the different mentalities compared with 20 or 30 years ago. He said that accidents can be divided into three major categories: disasters, dramatic injuries, and minor injuries and illnesses (Figure 3.1).

The first category, disasters, refers to accidents that can cause multiple fatalities. The second contains dramatic injuries caused by roof falls, inundation, and equipment accidents. The third category includes minor injuries and occupational illnesses such as knee and back injuries. According to this interviewee, the industry in the 1970s or 1980s, although it did not say so, accepted "a certain level of accidents," particularly minor injuries and occupational illnesses, believing that they were just part of the business. The industry focused only on disasters and dramatic injuries and did not think about how to stop those minor injuries and illnesses because they did not believe they could be stopped. For example, the company he worked for in the 1970s already had a safety program called Next Step to Zero. However, at that time, the program did not really look at all types

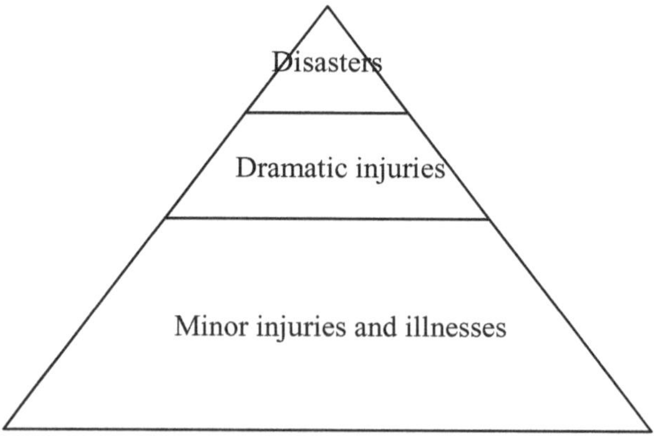

Figure 3.1 Three categories of accidents.

of accidents.[4] By contrast, the mentality of some of the largest companies today is that every accident is a problem and needs to be resolved, including minor injuries and illnesses.

There is no doubt that the change of mentality is an evolutionary process of continuous improvement in safety performance. Over the last few decades, the safety records of the coal mining industry, particularly some of the largest mining companies, have improved dramatically, resulting from the advance of technology and more systematic designing, engineering, and managing operations. For many years, the industry focused on the lost time injury frequency rate. Currently, in some of the largest coal mining companies, the lost time injury rate has been very low. For example, in Consol Energy, 12 operations did not have any lost time injuries in the first quarter of 2005. This number went up to 19 operations in the last quarter of 2007, accounting for more than 60% of its total operations (Holt, 2008).[5] As a result, the companies have seen the opportunity to eliminate all possible injuries and illnesses. In the words of an interviewee, "they are at the point now where they actually think to give a shot in getting to zero."

However, this change of mentality is also economically and socially driven. This has been demonstrated in both the scholarly literature and the information collected in interviews. Many informants indicated that running operations safely makes business sense. Mining companies today cannot "afford to have mining disasters" because "it affects the bottom line too much."[6] Safety makes business sense for at least three reasons. First, disasters or traumatic injuries may cause shutdowns of operations and loss of production. Today, "if you shut a longwall mine down, which is the most productive mine, you are probably talking about hundreds of thousands of dollars an hour of lost revenue." [7] Second, many of the largest coal mining companies today are publicly traded companies. Therefore, poor safety performance will increase the cost to access capital because their shareholders and lenders see the investment in unsafe operations as being riskier. Currently, in the annual report of almost every publicly traded mining company, workplace health and safety is always the first thing addressed. As an informant said, "25 years ago, safety was important (for the publicly traded companies). Now, it is absolutely No 1." Third, improved safety performance can also reduce the penalties paid for violations. This has become an increasing concern because the new legislation and regulations introduced after the Sago accident in 2006 have dramatically raised the civil penalties for violating the law. According to an interviewee, before 2006, the lowest penalty was $60. Now, it is $240 after the new legislation enacted in June 2006. Penalty amounts for some companies can go up 800 to 900%.[8]

However, according to informants, the change of mentality is driven not only by economic concerns, but also by social considerations such as enhancing reputations. Because of the significant improvement in safety performance, a disaster involving any fatality is highly visible and stigmatizes the company. In 2005, only 15 underground coal-related fatalities occurred in the entire year. There are 600 coal mines. If a coal mine has one fatality, it is in a group of 15 out of 600. It is a "clear exception to the vast majority of operations."[9] If a coal mine has multiple fatalities, there will be a lot of media and government scrutiny. Even worse, this stigma can never be "shaken off" because the media would bring this out again and again in the future when other accidents happen to occur.[10]

The social drivers are not totally separate from economic concerns. For example, improved reputation is also good for business because it can help to attract and retain employees. The lack of workers is one of the biggest challenges for the coal mining industry today. According to informants, the coal mining industry was very profitable in the 1970s, when a lot of people were being hired. However, the industry declined in the 1980s and 1990s, and there were not many job opportunities. Now the industry has become in demand again, but a lot of people who came during the 1970s are going to retire. Under this pressure, a good reputation can help a company to recruit young employees. As an informant stated,

> It also helps your recruitment if we go to a young mining engineer and say: "Look we are the safest mining company in the U.S. We care about our people." I think that is good selling point.

Again, the story described above illustrates the constitutive role of government regulations and civil society in the increase of industry self-regulation in the era of regulatory capitalism.

The drivers of mentality change are also the drivers for some coal mining companies to introduce the model of risk management. This is because, as previously stated, the search for new models is the logical next step when the mentality is changed. In the past decade, many coal mining companies have adopted various models to improve their safety performance. Some have established comprehensive safety management systems, such as Consol Energy. Others have embraced behavior-based safety programs. Also, very recently, some companies, such as Whitesburg Coal, have started to introduce the risk management model influenced by its worldwide diffusion. This will be addressed in the next sub-section.

As discussed in Chapter 2, on most occasions, people learn new ideas through modeling. What model and whether a model can be successfully adopted depend on a series of factors such as the attributes of observers (e.g., their social network and cognitive capacity) and the perceived properties of the model. Based on Rogers's (2003) theory of diffusion and Braithwaite and Drahos's (2000) theory of modeling, the next section will use Whitesburg Coal as an example to show how and why the company introduced the risk management model.

Modeling Risk Management in Whitesburg Coal[11]

At the time of research, only a handful of coal mining companies in the U.S. had adopted or were in the process of developing risk management systems. Some of them are parts of multinationals, which are very important platforms for diffusing models. However, models can also diffuse into domestic companies if at some point they connect with the source of models. In this research, Whitesburg Coal is selected as a case study to illustrate how the risk management model diffuses into American domestic coal mining companies.

During the time of the investigation, Whitesburg Coal was in the process of establishing risk management systems. The risk management model has been introduced in Whitesburg Coal for three reasons. First, the model mongers within the company have interpersonal connections with model missionaries and mercenaries. Consequently, Whitesburg Coal was exposed to the model of risk management. Second, the model mongers in the company have the cognitive capacity to perceive the value of the model. Third, the risk management model itself is compatible with Whitesburg Coal's existing management systems and culture. Each of these is elaborated below.

Model Mongers and Their Interpersonal Channels

According to the diffusion theory, people can acquire knowledge through mass media and interpersonal connections (Rogers, 2003). Although the mass media are the most efficient way to spread knowledge to people in widely dispersed locations, interpersonal contact is usually the more powerful trigger of modeling action. This is exactly what happened in Whitesburg Coal.

The model mongers in this case are safety professionals in top management because they are people who champion safety programs in the company. Without their support, nothing would happen. Two important figures regarding this issue are the Senior Vice-President of Safety Michael and the Vice-President of Safety John.[12] From what I learned in the field, Michael served as an industry representative to the International Labor Organization (ILO) in 2005. He met an Australian representative at an ILO conference. They spent a lot of time discussing the Australian model of risk management. When he returned, he continued this discussion by email about how to develop risk management systems. John joined the online discussion. These discussions through interpersonal channels triggered the action at Whitesburg Coal.

In late 2005, Whitesburg Coal initiated internal discussions on the next steps toward accident prevention. Michael and John introduced the issue about risk management, and everybody in the discussion felt that was the "key missing ingredient." Therefore, they decided to develop formal risk management systems in Whitesburg Coal. They did not encounter much political resistance within the company to launch this initiative because both Michael and John were on the top management team, the management had already established a strong commitment to safety, and most of the operations in the company were union-free.

Cognitive Capacity of Model Mongers

Exposure to a new model does not necessarily lead to change in behavior if the observer does not have the cognitive capacity to perceive the value of the model. The model of risk management has been introduced in Whitesburg Coal to a large extent because the model mongers, particularly John, saw the need to have risk management systems. John is a very important figure in this modeling process because he is the person who is in charge of developing and auditing risk management systems, as well as integrating risk management into existing management processes. According to the theory of sensemaking,[13] leaders are pivotal in an organization's cognitive process because they have the ability to influence others' sensemaking and meaning construction in an organization (Gioia and Chittipeddi, 1991; Weick, 1995).

During the time of this study, not many people in the U.S. coal mining industry understood the need for risk management. This is largely because they could not cognitively differentiate formal risk management from some informal risk management techniques that had been utilized in the industry for a long time. In practice, as described in Chapter 2, "risk management" has been used as a generic term, which can be understood as ranging from the most complex and structured notion of risk management to the simplest, involving risk identification and control for each task. As a respondent noted, risk management has "existed in the industry for a long period of time by different names and different levels of expertise."[14] Two commonly used informal risk management techniques in the industry are Job Safety Analysis (JSA) and behavior-based safety (BBS). JSA is an approach that allows people to analyze and manage risks inherent in an individual job, rather than system-wide issues. JSA has been used in the industry at least since the 1970s. Therefore, many people in the coal mining industry think that some of the processes they have been using are risk management. Actually, they are, but they are only some of the components of a formal risk management model. If people cannot differentiate advanced and comprehensive risk management, which starts from engineering the system-wide environment to mitigating risks and hazards in individuals' day-to-day performance, from informal risk management like JSA, they cannot see the need to have a formal risk management system. Based on my observation, John was one of the few people in the coal mining industry who understood the two different levels of issues at the time that the research was conducted.

For instance, at the Bluefield Coal Symposium, when John talked about risk management in his presentation, the organizer of the symposium who had been in the industry more than 30 years and who was a president of a major mining company, asked about the difference between risk management and JSA. From what he said in a following interview, I found that the organizer of the symposium basically saw risk management and JSA as the same, although he did to some extent understand that risk management is a boarder concept. He said:

> JSA has been used in the industry for years. It involves, basically, taking every job and breaking it down into tasks and then breaking it down into the hazards of those tasks and solutions—how to prevent these hazards from occurring. Risk assessment is a new word for JSA. *Today we talk about risk assessment. It is very much as same as JSA.* JSA involves every practical consideration of every machine. Risk assessment, I think, looks at wider experiences, company wide, long term as well as short term. I look forward to reading the NIOSH report to teach myself what is the difference between risk management and JSA. *JSA in my organizations worked very well. If they want to call it something else today, that is fine* [emphasis added].

John answered the question at the Symposium as follows:

> The level of hazard identification that individuals correct and deal with is behavior-based safety and JSA. Your action that needs a larger process and longer range is risk assessment. An analogy is roof control. The worker works on roof skin control, how to put a bolt in, does he put an extra bolt or a larger plate? That is his job; that is how you train him; that is JSA type of training. Risk assessment looks at the mine design, the block size, the major roof support system that was put in, do they work properly? If you use pillars for roof support at the tailgate, have they been tested? That is the design for your type of mine and not taken from someone else. Years ago, that is just how we did it. You take something from someone else. It worked there and it will work here. Risk assessment says you don't do that.

Based on this understanding, Whitesburg Coal, led by John and Michael, decided to develop both BBS and risk management at the same time. In John's words, they set out to "push the behavior-based [approach] up and risk management down." In other words, they saw risk management as a different tool from BBS. The former is the management's tool used to assess the system-wide environment to prevent catastrophic risks that can cause multiple fatalities; whereas the latter is the workers' tool to deal with day-to-day job failures. John explained:

> From my view, there is an old cliché: if the only tool you have is a hammer, all you should use is this. I am going to hit them all. That is

> the only thing in your tool box. . . . My concern is when you say you brought a number of tools in your tool box and use them all. . . . One of the things we are talking about is how do you blend the scene. I am saying you start with, relatively speaking, all managers. You have a small group to get risk assessment started with, engineering, safety, some of your operational arms. You don't need to get down to the foreman level. The foreman level is hazard identification [BBS]. . . . That [risk management] is you look at multiple fatality disasters that you are trying to assess. You are not trying to assess lifting, pose, and how the guy gets hurt on his back. That is a different level of issue.

The reason that John can perceive the need for risk management is to some extent because of his own experience. As previously described, people learn not only through modeling but also through enactive learning based on direct experience. According to Bandura's social cognitive theory (1986, 111, 116) that stresses reciprocal determinants among environment, individual psychology, and behavior, the environment influences people's behavior through the "intervening influence" of human thought that is of an "abstractive, reflective, and generative nature." In other words, people constantly verify their behaviors by drawing generative lessons from their previous actions.

During the interview, John told me the story about his cognitive evolution in terms of the need for risk assessment. It started from a disaster he was involved with in his previous workplace in 1988. At that time, there was an operational issue of coal quality in the Marianna Mine No. 58 in Washington County, Pennsylvania. In order to keep the mine open, the whole mining plan needed to be changed to access the area where the coal quality was better. The company put in some sections that allowed workers to mine coal in one area rather than in different spread areas. This change required more ventilation. They added a larger fan, which dramatically increased the fan pressures; turned belt entry into intake; and put in an atmosphere monitoring system. All these changes were undertaken without going through any risk analysis. John said that, at that time, they did not know they needed to do risk analysis. They had thought that all these changes were done safely.

Unfortunately, on March 7, 1988, a fire started at a belt drive. According to MSHA (2008a), belt slippage and frictional heating ignited the loose coal that had accumulated in the drive rollers. The fire water system did not put it out. The fire "quickly propagated down the belt, ignited other combustibles, and totally engulfed parts of the belt entry." About 20 miners in working sections were evacuated. Five of them were sent to a hospital for smoke inhalation.

People in the company reviewed the event and analyzed what caused the accident. According to John, all the analysis they did started from the day of the event and traced back. Later, when he had a chance to think about it, he recognized that the problem actually started long before the event. It started when they decided to change the mine design. He started to believe that any change needs to be planned through proactively. It should not be done after the fact, which is what the government does. At the same time, he was taking a graduate course on risk. When he read about risk, he thought "aha, this is really what we should be doing." During the interview, he said that if they had done a risk analysis before they changed the mine design, they would either have chosen not to do it at all or tried to minimize the risk to prevent the fire from happening.

Although Whitesburg Coal has been exposed to the risk management model through interpersonal channels and the modal mongers in the company also have the cognitive capacities to understand the value of the model, the modeling process would have been very difficult if the model did not fit well with Whitesburg Coal's existing culture and systems. According to Rogers (2003), the less a model is perceived as being consistent with existing values and past experiences, the less likely it is to be adopted. Fortunately, the risk management model is compatible with Whitesburg Coal's existing culture and institutional arrangements.

Compatibility

Before Whitesburg Coal decided to introduce the risk management model, there were already some management systems in place involving informal risk assessment for stand-alone tasks. These management systems, which emphasize structure and planning, provide the basis to integrate a risk management model into the existing management systems.

According to John, many management systems and people on the management team of Whitesburg Coal came from its predecessor company, CA, which had a strong tradition of risk assessment and management.[15] Some safety management systems of CA involved elements of risk assessment, both formal and informal. For example, CA had a formal program called Quest 21, which was designed to allocate resources according to the level of risk. In other words, more money was spent on the issues having a higher level of risk. This program is called the Cost of Quality at Whitesburg Coal. Although the terminology is different, the basic structure of Cost of Quality is as same as that of Quest 21 in CA.

Besides the Cost of Quality, some other management processes in Whitesburg Coal also involve risk assessment. One of them is Search for Excellence, which is part of a continuous improvement process. In this process, each operation uses a local team to identify issues that need improvement. Search for Excellence teams usually use risk assessment techniques, particularly when anything needs to be changed, such as mining plans or longwall moves. For example, a mine has a ventilation plan for a certain number of years. Whenever the mining plan changes, risk assessment is used to determine if the original ventilation plan still works and if another shaft, more entries or more cushions are needed.

Because of the existence of these management processes in the company, people at Whitesburg Coal perceive the risk management model as having less complexity than do those at other companies. John cognitively interprets the risk management model as a "logical connection" to the Search for Excellence process because "both processes are designed to look up further into the future and both used similar tools to identify and prioritize risks."[16]

The existence of these management processes also makes it much easier for them to integrate the risk management model into their existing safety management systems. According to the theory of diffusion, a model has to be re-invented to fit into an adopter's existing culture and structure (Rogers, 2003). This is exactly how Whitesburg Coal introduced the risk management model. When John described his responsibilities, he said:

> The other thing I have to do is to make this a natural fit into our already established management system . . . you cannot have a safety process like the risk assessment sitting over here and say that is a separate part of project. It has got to fit some place in the systems. My role will be to make sure it fits in the system and that it is not outside of the normal way we do business.[17]

In sum, Whitesburg Coal was trying to integrate different existing pieces of process, such as Cost of Quality and Search for Excellence, into an umbrella risk management program. This integrated program involves more formal requirements for conducting risk analysis before performing certain tasks.[18]

In addition, the management culture in Whitesburg Coal is formal and structured partly because it owns many large mining operations. Therefore, people in Whitesburg Coal do not feel uncomfortable when they are exposed to the highly structured and paper-oriented risk management model. An informant said:

> [Whitesburg Coal]'s culture is very rigorous, and paper and documentation oriented. . . . We think we already passed that hurdle. . . . We feel we are ready, and it won't be a huge change for us.

Because of the reasons discussed above, the risk management model has been successfully introduced into Whitesburg Coal. Unfortunately, although the management decided to develop more formal risk management systems in late 2005, progress was much slower than expected. The major reason is that the company has been fully occupied by dealing with new government regulations after the Sago accident that occurred on January 2, 2006. In John's words, the company "lost 2006, 2007, and 2008." The negative influence of government regulations on its modeling process will be further discussed in the next chapter.

Despite the delay, he said that Whitesburg Coal may be "better off in the long run," because safety professionals were exposed to more information about the Australian model of risk management, resulting from some efforts taken in the industry to introduce the model after the Sago disaster. Therefore, he said that the model Whitesburg Coal "puts together may be better because the learning curve would be steeper" if it had done it by itself.[19] According to him, the participation in the National Institute for Occupational Safety and Health (NIOSH) risk assessment project was the biggest jump for him and his staff in terms of understanding risk management. This indicates that modeling may be more likely to succeed if the company can get help from the collective efforts of the industry, although firm-to-firm modeling is also possible. The next two sections address the process of modeling risk management in the American coal mining industry and its impact on individual companies.

Modeling Risk Management in the American Coal Mining Industry

Change of Mentality—Emerging Common Fate

As shown in the analytical framework in Figure 2.5, the bottom-up process of driving up standards at the enterprise and industry levels is a proactive process starting from individual entrepreneurs who promote regulatory innovations. By contrast, the top-down change of regulatory policy is more likely to be a reactive process in response to disasters that provoke the mass public. However, based on my observation in this case, driving up standards at the industry level can also be a reactive process to disasters. As the previous section described, some of the largest coal mining companies have proactively changed their mentality and adopted various safety management systems to improve their safety performance. Nevertheless, until a series of accidents occurred in 2006 and 2007, they did not have strong motivations to drive up standards across the industry.

When talking with respondents about why 2006 and 2007 made a difference, the things they mentioned most were the media and public perceptions. In general, people's perception of the mining industry is very negative probably because of its historically poor health, safety, and environmental performance. In the eyes of many people, coal mining is "dark, dirty, and dangerous" (Harvey, 2007). Although the safety performance of the mining industry has improved significantly over the past several decades, public perceptions have not changed much.[20] This is partly because human nature tends to focus more on bad things. As an informant said, "good things that happen don't really make good news . . . the bad things make the headline."

This negative public perception is further reinforced by the mass media. The modern media and mass communications make disasters more visible to more people. According to an informant who has been in the industry more than 20 years, he was "shocked by the number of media" and the intensity of coverage when he was on the site of the Crandall Canyon disaster in 2007. When some other interviewees talked about recent disasters—Quecreek in 2002, Sago in 2006, and Crandall Canyon in 2007, they said that the media was on the scene within one or two hours after the events and the media coverage was "minute by minute and hour by hour." Even worse, in cases where miners were missing, the media focused on the accident for days and even weeks. For this reason, mining accidents are more likely to draw media attention than other events. An informant described:

> I think one of the reasons mining disasters got so much attention is because people are missing, people are trapped, and you don't know their condition, whether they are alive or not. It draws a lot of interest . . . if you have a plane crash and 200 passengers are killed. It is on the news. The next day they still talk about it a little bit and the third day you may hear one comment. In the mining industry, like the miners in Crandall Canyon in Utah, they were missing for 10 days and people kept saying we think they are alive, so the attention grows. You say in a minute, we are going to find them. That is probably unique to mines. If you remember the Quecreek accident in Pennsylvania, those miners were missing for three days and all of sudden they were brought out alive, so much media attention arises because it is almost like a drama. Human nature is to be curious about it.

Because of the negative public perception and increasing media scrutiny, the largest coal operators in the industry have recognized a common fate for the whole industry. Currently, "if something happens over here, it affects the guy over there."[21] This is because, first, public perception often drives the governmental and public action against the industry's role as the key provider of energy. They may call for the use of alternative energy sources such as wind and nuclear energy.[22] Second, the poor public perception can lead to a negative regulatory climate for the industry because the operators' views tend to be ignored during the legislative review process (Harvey, 2007). These two points are exactly the messages that Brett Harvey, the CEO of Consol Energy, sent to the industry in his famous speech in 2007, which will be further addressed later. Third, a poor image of the industry will affect the recruitment of employees. As mentioned earlier, the shortage of employees is one of the biggest challenges for the industry right now. During an interview, an informant said:

> You are going to go out to get a job somewhere after your PhD. I don't know what your plans are, but the coal industry would probably not be one of your first choices. If you go back to your graduate groups, how many guys want to work in the coal industry? I would imagine all would look at you and say "what you are talking about?" But the per-

ception of your fellows, even at Virginia Tech, which has strong mining history, the majority of students . . . if you did a poll down there . . . won't even think about working in coal companies, they think we are bad people. That is what I am saying; you cannot get the brightest into an industry that is not considered to be a good industry to work in.

As a result of the recognized common fate, some early movers—the largest mining companies that have built up commitment to higher level safety standards—have initiated actions to drive up standards collectively in the whole industry, led by individual entrepreneurs such as CEOs and senior safety professionals.

Driving up Safety Standards in the Industry

Some of the largest coal mining companies have taken two types of actions. First, they have publicly called for a change of mentality in the entire industry. Second, they have engaged organizational powers to search for concrete mechanisms to drive up standards.

As previously stated, although the early movers have changed their mentality for many years, they rarely spoke publicly until they saw the need to change the mentality for the industry as a whole. This change happened after the Crandall Canyon disaster that occurred on August 6, 2007, in Utah. The first to call for a change of mentality in the coal mining industry was Brett Harvey, the CEO of Consol Energy. Just 17 days after the Crandall Canyon disaster, he gave a famous keynote speech at the Utah Mining Association's 92nd annual meeting. In his speech, he first warned the industry that its poor reputation threatened the future of the industry. He said that the "blows to our industry's reputation" not only impacted the safety regulatory environment, but also jeopardized the role of the industry as a major source of energy. He asked, "If we are perceived as 'a dirty business' when it comes to safety, why would our friends in Congress or the agencies work with us on other important issues?" (Harvey, 2007).

Then, he pointed out that "the best way for the industry to close the reputation-reality gap is to eliminate accidents . . . to be at zero." He called for the industry to drop the traditional incremental approach because those who are satisfied with incremental improvements in safety actually tolerate a certain level of accidents.

> Our incident rate last year was less than 3.00 company-wide. We were very proud of that record. But, in human terms, it equated to 236 individuals being injured out of 7,500 employees. So if we improve our safety record this year by 20%, it still means that 189 of our people will get hurt this year. If we accept this incremental approach, we had better be prepared to ask for the 189 volunteers who want to get hurt this year. (Harvey, 2007)

He actually sent the message to the industry that simply reducing accidents is unacceptable and the only acceptable rate is zero. At the end of his remarks, Harvey said:

> We need to change the paradigm and we need to change it now. There is no one sitting in this room today who wants to see an employee get hurt. Our past is recorded and set. The future of safety is ours to decide. What industry must change is our incremental approach to safety improvement because it creates an unintended level of tolerance to accidents. We need to get to zero. Let's get to zero NOW!

Here we can see again how the shadow of government regulation constitutes the mentality in the industry. Brett Harvey's speech was very influential in the industry. Almost every informant at the management level mentioned this speech. According to a vice-president of safety at a mining company, his CEO heard this speech and sent it to him. Quoting his words, this speech "sets the tone" for the industry.[23]

Following Harvey's speech, another influential event that demonstrates the change of mentality in the industry was the Bluefield Coal Symposium in August 2008. The topic of the symposium was "Coal Mine Safety: the Road to Zero Harm." Around 250 people came to Bluefield, a small town in West Virginia. The participants included CEOs from almost all of the biggest Eastern coal operators.

In addition to calling for a change of mentality, some of the largest mining companies took another step by creating an organizational platform to figure out specific ways to drive up safety standards. One such platform is the independent Mine Safety Technology and Training Commission sponsored by the National Mining Association (NMA) right after the Sago accident in 2006. The decision was made by the NMA's board of directors, which consists of the CEOs of most of the largest mining companies in the U.S. When talking about the purpose of this commission, an interviewee said:

> We were doing very well prior to January 2006. We had seen continued improvement of safety and health both in terms of the number of fatalities, injuries, and accidents. On January 2, 2006 (the day of Sago accident), we woke up and it was a new day and a new year. We needed to start taking some fresh looks and thinking outside of the box. We needed to see if there were things we weren't doing that we needed to be to continue the downward trend and get back on that path. That is why we appointed the commission. We wanted people to take a fresh look, without being biased by part of the association.

The NMA appointed Dr. Larry Grayson as the chairman of the commission. He was the chair of the Department of Mining and Nuclear Engineering at the University of Missouri—Rolla at the time. Although NMA provided the financial support, it was Grayson who chose the members of the commission. Grayson put together a multipartite commission including all major stakeholders of the industry. There were representatives of operators (e.g., Brett Harvey, CEO of Consol Energy, and Anthony Bumbico, Vice-President of Safety for Arch Coal), unions (Cecil Roberts, the president of the United Mine Workers of America), a government research institute (Jeffrey Kohler, Associate Director for Mining and Construction at NIOSH), and universities (Thomas Novak, Chair of the Department of Mining and Mineral Engineering, Virginia Tech). The commission also included a member from outside the mining industry—Dr. Amy Donahue, Associate Professor of Public Policy at the University of Connecticut. In December 2006, the commission issued its report, *Improving Mine Safety Technology and Training: Establishing U.S. Global Leadership*. The report made 71 recommendations. The first recommendation of this report is a paradigm shift focusing on accident prevention and the establishment of systematic risk-based management systems.

Meanwhile, there have been some informal discussions among some safety professionals of the largest coal producers on how to "put a different face on the industry" by creating some mechanisms to separate good performers from bad ones.[24] An informant recalled that, when he discussed the pending Supplementary Miner Act (S-Miner Act) in Congress after the Crandall Canyon disaster with some colleagues from the industry, a vice-president of safety from a mining company said:

> We need to have some sort of ways to separate some operators from the others. If you meet these standards, if you do these things, you are in this group. If you don't want to meet these standards, it is not to say you are illegal; you comply with rules, but you are in a different category.[25]

A typical example of this kind of mechanism is INPO in the nuclear industry, which is an industry self-regulatory body to write and enforce a standard of excellence beyond compliance with minimum standards in laws.

All of the CEOs and vice-presidents of safety mentioned above are model mongers. They create the opportunity to adopt new models, but they may embrace different models to see which one works best. When the model mongers have created the platform, model missionaries and mercenaries come in to sell the model that they believe is the best. The next section will show how the industry collectively models risk management, particularly the role of model missionaries and mercenaries in this process.

Modeling Risk Management in the Industry

As previously described, the report of the Mine Safety Technology and Training Commission chaired by Grayson recommended that the industry adopt risk-based management systems focusing on major hazards. This is the first formal and documented introduction to risk management in the industry. Therefore, it can also be seen as a starting point of the modeling process in the American coal mining industry. The modeling process could begin for two reasons. First, as mentioned earlier, the model mongers—the CEOs in NMA—had the foresight to set up an organizational platform, the Mine Safety Technology and Training Commission, and also allowed the Commission to work independently. Second, when the platform was ready, model missionaries stepped on the platform and introduced the model of risk management.[26]

Grayson was the most important model missionary in this process. He was the one who brought up the issue of risk management to the Commission. A model missionary is a person who advocates a model not only for financial reasons, but also for the reason of belief. Grayson was a mine superintendent before he came into academia. He said that he firmly believes that mines should proactively identify and mitigate risks by looking at historical data of violations and citations. In recent years, he has produced many publications and reports on risk analysis and the culture of prevention. In addition, he can become a model missionary on risk management because of his cognitive patterns. He says that he "tends to be a data person" and likes to use risk analysis to see where he stands and to set goals. This is perhaps because he holds his first bachelor's degree in mathematics. According to him, the first time he used risk analysis was to prioritize research when he was the associate director of the NIOSH office for mine safety and health research about ten years ago. He formed a team of researchers who developed a risk matrix based on the frequency and consequence of various incidents that had occurred over the last 30 years, including dust, noise, roof falls, explosions, fires, and hits by equipment. According to the ranking in the risk matrix, he determined what proportional amounts of resources needed to be allocated to each area.

Grayson had several channels to spread the idea of risk management to the industry. First, he attended several meetings to disseminate the recommendations of the Commission report, including the idea of risk management. He gave an executive briefing at the NMA board of directors' meeting and in its health and safety committee regular meetings. He also attended the Bluefield Coal Symposium in 2007 and spent about two hours going through the recommendations.[27] Second, during the time of the research, he also tried to bring some companies into his research using a system safety analysis approach.

Although the Commission report drew industry attention to risk management, it did not give a thorough explanation of what risk management is and how to conduct risk assessment. Therefore, considering the lack of understanding of risk management in the industry, the Commission recommended that NIOSH develop templates that can be used for various types of mines. In fact, NIOSH initiated a project of developing risk assessment templates almost at the same time as the Commission was formed, because the leadership in NIOSH and some researchers saw the opportunity to develop knowledge that the industry did not have.[28] Assisted by NMA, they persuaded Congress to allocate more funding for mine safety and health research after the Sago accident. The study of risk management and assessment is one of the projects Congress funded. Therefore, the researchers in NIOSH are another group of model missionaries. The key figure is Anthony Iannacchione, who is the primary author of the project's final report, *The Application of Major Hazard Risk Assessment (MHRA) to Eliminate Multiple Fatality Occurrences in the U.S. Minerals Industry*, which was published in October 2008. He went to Australia to take courses on risk management.[29]

In addition to these model missionaries and mercenaries in the U.S., others from the source of the model in Australia were also involved in this modeling process. The most important one is Jim Joy, a well-known Australian scholar and consultant on risk management in the mining industry. According to Joy, he was invited to talk about risk management at the Academy of Science in Washington, DC, after the Sago accident. That is where he met Jeffery Kohler, the associate director of the Mine Health and Safety Department at NIOSH, who initiated the idea of doing a risk assessment project for the American mining industry. When the NIOSH MHRA project was launched, Joy was invited as the facilitator of the study.

Joy started as a risk facilitator in nuclear and conventional power plants in Canada. He immigrated to Australia in 1996, where he became involved in risk management in the mining industry. Sponsored by several companies and the government of Queensland, he created the Minerals Industry Safety & Health Centre (MISHC) at the University of Queensland in 1998. MISHC developed a Minerals Industry Risk Management (MIRM) model in 2006. This model has been adopted by most Australian mining companies. The method of the NIOSH MHRA project followed almost exactly the MIRM model.

In 2006, Joy held a series of workshops at NIOSH. The industry was encouraged to participate. Some informants, such as the Vice-President of Safety at Whitesburg Coal, participated. The MHRA research team asked participating companies to choose one major hazard as a pilot case study. Ten companies were picked for the pilot studies. These companies came from different sections of the industry, ranging from small to large size and from coal to non-coal mines.

The studies focused on major hazards that may cause multiple fatalities. According to the Grayson commission recommendations, risk management of these major hazards is the first step that the industry needs to focus on. The coal mine case studies in this project focused on five major hazards: spontaneous combustion, underground workshop fires, water inundation, conveyor belt fires, and longwall gate entry track fires (Iannacchione, Varley, and Brady, 2008). The risk assessments in the ten pilot case studies were conducted by the personnel at the mine sites. Joy and NIOSH experts came to the sites to provide training and facilitation. Therefore, the process of doing the studies themselves was a process of disseminating knowledge about risk assessment and management.

The final report of the study was published in October 2008. The most enlightening part of the report was the introduction of basic techniques and the process of risk assessment and management. In terms of risk management techniques, it clarifies the different levels of technique, ranging from informal to basic formal and advanced formal risk management techniques. The informal techniques focus on worker behavior. A typical example is the Stop-Look-Analyze-Manage (SLAM) process that has been adopted by many companies in the U.S. coal mining industry. As a basic formal risk assessment technique, Job Safety Analysis (JSA) is the best example. JSA requires workers and management to develop structured and documented standard operating procedures to eliminate or reduce potential hazards associated with each task.

The U.S. coal mining industry is familiar with these two risk assessment techniques. What they are unfamiliar with is the third technique, which is called "advanced formal risk management" in the NIOSH report. This technique requires the "use of a structured approach that incorporates one or more risk analysis tools and produces a documented assessment of the risk associated with unwanted events" (Iannacchione, Varley, and Brady, 2008, 10).

Moreover, this report gave detailed explanations of different commonly used risk analysis tools including Workplace Risk Assessment and Control (WRAC), Preliminary Hazard Analysis (PHA), and Bow Tie Analysis (BTA) described in Chapter 2. In addition, this report outlined the typical process of qualitative risk analysis, which goes through the following steps:

- Risk Assessment Design—determining target hazards, time framework, and tools to be adopted;
- Risk Assessment Team—selecting team members;
- Risk Assessment—conducting risk analysis including identifying risks, ranking risks, identifying existing and new prevention and recovery measures;
- Effectiveness of Controls—assessing the effectiveness of identified measures following the hierarchy of control framework;
- Audit and Review—determining what control measures to be adopted and assigning responsibilities for the action to be taken (Iannacchione, Varley, and Brady, 2008, 17–22).

The NIOSH project on MHRA is the most important step of modeling risk management in the U.S. coal mining industry. The report is actually a product of symbolic interpretation. It can be transmitted to the wider community and guide future actions. More important, this report may improve people's cognitive capacity by clarifying the different levels of risk management and their functions. As previously mentioned, the lack of cognitive capacity to differentiate advanced risk management from informal risk management techniques prevents many in the industry from recognizing the value of the risk management model.

Since the lack of understanding is one of the biggest hurdles for the American coal mining industry in embracing the risk management model, the model missionaries and mercenaries who possess knowledge of the model play significant roles in the modeling process. For example, in the acknowledgments of the final report of NIOSH MHRA project, the authors wrote:

> We wish to recognize the guidance and efforts of Professor Jim Joy in facilitating the MRHAs. He was forever teaching and all were his interested students. (Iannacchione, Varley, and Brady, 2008, 3)

These model missionaries and mercenaries as well as the knowledge-based organization supporting them together have formed an epistemic community, a hub for diffusing the model of risk management in the U.S. After the report was published, NIOSH and NMA planned to hold workshops to disseminate the methods of risk management across the industry. This important role of model missionaries and mercenaries echoes one of Levi-Faur's (2005, 27) major characteristics of regulatory capitalism—"the growth in the influence of experts in general and of international networks of experts in particular."

After having discussed the modeling process of risk management in the industry, the question arises of how this modeling process has affected the modeling processes in individual companies. This is the question the next section examines, again using the example of Whitesburg Coal.

Impact of Modeling in the Industry on Modeling in Companies

Since modeling happens at different levels and in different entities, different modeling processes can influence each other. A successful modeling of a person or an entity can stimulate and facilitate modeling of others (Bandura, 1986, 50).

The influences on the modeling process in Whitesburg Coal are primarily from the model missionaries and mercenaries and the organizations behind them such as NIOSH. According to John, the Vice-President of Safety at Whitesburg Coal, the recommendation of the Grayson commission drew his attention to systematic risk management. He had known Grayson for many years. When Grayson was a mine superintendent, John was the safety director in a neighboring mine.

As John stated, participation in the NIOSH project was a big jump in modeling advanced risk management in his company. This is because, through participation in this project, he and his staff were exposed to more information about the model because of the direct contact with model missionaries and mercenaries; they also had the opportunity to learn by doing. John joined the NIOSH workshop on risk management in 2006, where he met Jim Joy from Australia. John decided to participate in the project, and they chose one large longwall mine with more than 600 employees as a pilot case study. John, Joy, and experts from NIOSH went to the mine site to help the local team conduct the risk assessment. According to the information from the final report, the team was made up of four representatives from the company's safety program, a general maintenance mine foreman, a continuous improvement coordinator, a fire brigade supervisor, two external fire prevention experts, a NIOSH observer and a facilitator (Jim Joy). The risk assessment is considered a success in the report. As an interviewee who was a team member said:

> It opened our eyes about the usefulness of the tools. Through doing the exercise we realized that the better fit for us would be to use it on a higher level as an overall culture.

CONCLUSION

Based on the discussion above, it can be seen that the rise of risk management in the American coal mining industry is not an isolated process. Rather, it is embedded in a larger picture—the industry's collective process of driving up standards. The process of driving up standards starts from the change of mentality in some of the largest mining companies. Led by individual entrepreneurs—CEOs and senior safety professionals—these companies have established commitments to higher safety standards. When these individual entrepreneurs recognize a common fate among industry actors because of the pressures imposed by government regulations, civil society, and the public, they call for a change of mentality in the industry as a whole by upholding higher level standards, not only for economic reasons, but also for reputational concerns. They also actively create organizational platforms to search for models to fulfill their commitments to higher level standards.

This process of driving up standards in the American coal mining industry is largely consistent with the expectation of the analytical framework. However, one modification is needed. According to Braithwaite and Drahos (2000), there are both proactive and reactive sequences of regulatory change. The former is a bottom-up process, which starts from individual entrepreneurs in the industry who promote regulatory innovations, whereas the latter is a top-down process when the state realizes the need for action in response to disasters. This study suggests, even in the proactive process of driving up standards in the industry, there was a reactive logic. In response to a series of disasters that occurred in 2006 and 2007, individual entrepreneurs in the American coal mining industry recognized the common fate and the need to drive up standards across the industry.

The change of mentality and the creation of organizational platforms by individual entrepreneurs provide the conditions for a modeling process to start. Nevertheless, other conditions are needed for modeling a particular model like risk management to occur. First, there have to be model missionaries and mercenaries who spread the model based on existing or newly created institutional platforms. Larry Grayson, who brought out the risk management idea in the Mine Safety Technology and Training Commission sponsored by NMA, and Jim Joy who provided training to NIOSH researchers and the American coal mining industry in the MHRA project, are the examples in this case. Second, on the demand side, there must be model mongers who have both a connection to the source of the model and the cognitive capacity to understand the value of the model. John in Whitesburg Coal is a model monger of this sort. In addition, a model is more likely to be adopted in an organization if the model is perceived to be compatible with existing organizational structures and culture. This is exactly what happened in Whitesburg Coal. Therefore, the existence of all these conditions explains how and why the risk management model has been introduced in the coal mining industry in the U.S.

Despite the progress in modeling risk management in companies and the industry as a whole in the U.S., the industry is still at a very early stage of the modeling process. The movement is not as quick as many people expected. Neither the Grayson commission's recommendations nor NIOSH's MHRA report can impose much pressure on companies. The extent to which the recommendations and risk management methods will be adopted depends completely on the companies.

An informant portrayed a possible three-step journey of driving up safety standards by embracing risk management. He said that the Grayson commission report is really the first step because it is the first written document discussing the systematic risk management approach. The second step is a discussion about how to do it. He said that the conversation about the mechanisms to separate the good from the bad companies is "at best the second step." He also hoped this step would be taken in Joy's workshop at Minexpo in Las Vegas a month after the interview. In the third step, he expected some companies that are working on risk management would say, "ok, we need to work together to make ourselves stronger . . . we need not reinvent the wheel each time."

Despite his expectations, however, only about a dozen persons coming from three to four companies attended Joy's half-day workshop at the Minexpo. The workshop was like an introductory course. There was no discussion about how to implement risk management in the U.S. Moreover, by the time this research ended, the NIOSH MHRA report had had little evident influence. Almost a year after its publication, no dissemination workshops have been organized,[30] and even some interviewees who are important industry people are not aware of the NIOSH report. The question then arises: why has the model of risk management not been as popular as some people expected in the U.S.? This is the question later chapters will address.

4 Barriers to Modeling the Risk Management Approach in the American Coal Mining Industry

INTRODUCTION

The last chapter showed that, although risk management has been introduced in the American coal mining industry to some extent, the diffusion process has been very slow. This chapter seeks to answer the second subquestion: what are the barriers keeping the coal mining industry from more fully embracing the risk management model? A comparative perspective is adopted to examine this question. First, the U.S. coal mining industry is compared with Australia's where risk management has been used extensively. Second, since the majority of American coal mining companies have not introduced the risk management model, it is worth exploring the barriers that keep them from embracing this model. Blacksburg Coal, an example of companies of this sort, is compared with Whitesburg Coal, which has introduced the model.[1]

The analytical framework suggests that some structural variables can be barriers to the modeling process. This research focuses on two such variables: the characteristics of the industry and the regulatory environment. I find that the existence of a large number of small operators is a major barrier keeping the industry from embracing risk management. However, the diversity of the industry structure is partly because of weak international competition and relatively lesser importance of the industry in the national economy. As a result, the U.S. coal mining industry has not undergone large-scale consolidation as has its Australian counterpart. There is also less feeling of common fate, and, therefore, fewer reputational concerns and less motivation to adopt best practices such as risk management in the industry. The second barrier is highly prescriptive regulation, which consumes the resources that companies could use to develop a risk management model and creates a mentality of compliance that is not compatible with the idea of risk management. In addition to these structural barriers, the research reveals that many American coal mining companies find it easier to accept a competing model of risk management—behavior-based safety (BBS). The rest of this chapter will address these three barriers.

CHARACTERISTICS OF THE INDUSTRY

At first glance, the major difference between the American and Australian coal mining industries is the different industrial structure. There are a large number of small operators in the U.S. but not in Australia. However, this is not the fundamental difference. Rather, the fundamental differences are twofold. The first is the different levels of connection to the international market. More specifically, the Australian coal mining industry is export-oriented, whereas the American coal products are consumed predominately in the U.S. According to EIA (2009d), coal exports in 2008 reached the highest level in more than a decade in the U.S. In this year, 81.5 million short tons of coal were exported, accounting for 7% of the total production (1,171.5 million short tons). By contrast, Australia exports approximately 60% of its annual coal production (EIA, 2009c). Australia is the fourth biggest coal producer in the world, but it is the largest coal exporter, accounting for 29% of global coal exports (*ibid.*) (Figures 4.1 and 4.2).

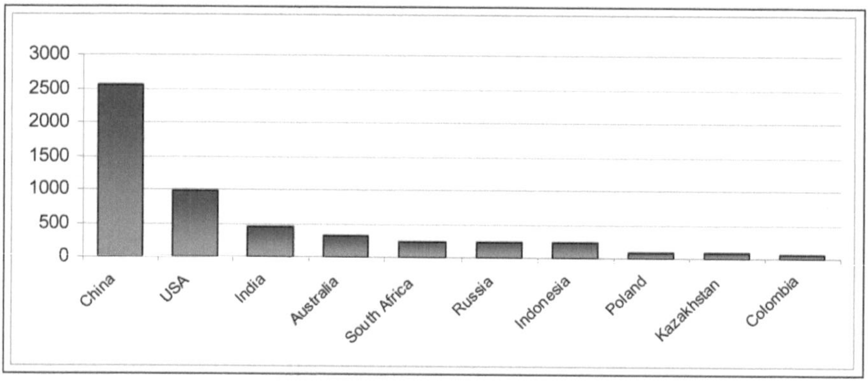

Figure 4.1 Top coal producers, 2007 (mt).
Source: World Coal Institute (2009a).

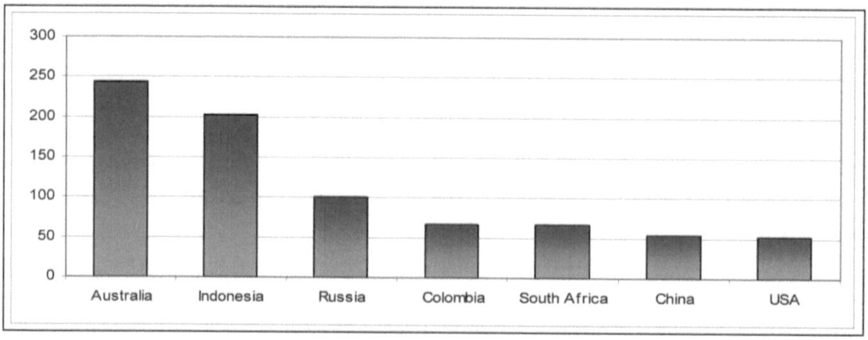

Figure 4.2 Top coal exporter, 2007 (mt).
Source: World Coal Institute (2009b).

Table 4.1 States in Which Mining Generates More Than 1% of GDP

States	GDP Generated by Mining (%)
Wyoming	8.1
West Virginia	5.9
Montana	2.7
Kentucky	2.1
Nevada	2.1
Alaska	1.6
Arizona	1.6
Utah	1.3

Source: NMA (2009a).

The second fundamental difference is the different level of importance of the industry to the nation's economy. Unlike many other developed countries, the Australian economy depends much more heavily on the mining and minerals sector. The exports of the sector account for almost half of all exports of the country (Sheehy and Dickie, 2002). The sector's contribution to gross domestic product (GDP) has been around 9% (Starke, 2003, 45). By contrast, less than 0.5% of the GDP comes from mining in the U.S. (*ibid.*). According to NMA (2009a), mining's contribution to GDP was no more than 1% in most states in 2007. Only in the states in Table 4.1, mining generates more than 1% of the state's GDP.

Because of its strong reliance on exports, the Australian mining industry faces increasingly tough competition in the world market. On the one hand, the growing economies in Asia such as China dramatically increase the demand for minerals. On the other hand, the major export destinations for Australian mineral products are Asian countries, such as Japan, South Korea, China, and India (EIA, 2009c). Therefore, the industry encounters tough competition from low-cost producers in the Asian and Pacific regions, such as Indonesia and China. One result of this stiff international competition on both the demand and supply sides was a large-scale consolidation in the 1990s to increase efficiency.[2] As a consequence, almost all coal mining companies in Australia became large corporations.

Added to the importance of the sector to the economy, Australia is more vulnerable to fundamental threats to the mining industry. As such, there is a strong feeling of common fate not only among all operators in the Australian mining industry, but also among all stakeholders: operators, governments, and labors. All of the parties have a stake in the prosperity of the mining sector. The decline of the mining sector would have a significant impact on the national economy as well as on the wellbeing of miners. For example, the website of the Australian Department of Resources, Energy and Tourism includes phrases such as "enhancing Australia's economic prosperity" and "the Australian Government is committed to creating a policy framework to expand Australia's resource base, increase the international competitiveness of our resources sector." These words illustrate the government's strong interest in the success of the industry.

The consequences of the strong common fate are twofold. First, operators and major stakeholders are bonded together. The industry is like a community. There is a high level of cooperation and dialogue in the industry. This issue will be discussed further in the next chapter. Second, there are more reputational concerns in the Australian mining industry. The Australian mining industry relies heavily not only on exporting mineral commodities to other countries, but also on having access to mineral resources in the world. Without a good reputation, the industry will face more counteractions against their operations by local communities, NGOs, and local governments. In this environment, a good reputation is crucial for the success of the sector. Therefore, the Australian mining industry is keen to adopt best practices to build its reputational asset. As Braithwaite and Drahos (2000) note, a regulatory community tends to embrace principles or best practices that can enhance its reputation when a common fate is recognized. For example, the NSW Mine Safety Advisory Council (MSAC), a major tripartite advisory body to the Minister of the Department of Primary Industries in NSW, appointed Norman Jennings as the independent chairman in 2006. Jennings was the former director of the mining division at the International Labor Organization. He said that the objective of MSAC is to implement the Wran Mine Safety Review recommendations, and then "demonstrate the extent to which world best practice in mine health and safety is used throughout NSW, thus setting the scene for the next phase of its life" (Jennings, 2009). It can be seen that the world's best practices is the benchmark of the NSW mining industry in Australia.

Compared with its Australian counterpart, the American coal mining industry is less internationally oriented and less important to the national economy. Consequently, the American coal mining industry has less feeling of common fate, and, therefore, has weak reputational concerns and is less motivated to adopt world best practices, such as risk management.

In addition, U.S. mining companies did not go through the large-scale consolidation that happened in Australia, partly because of less competitive pressures from the international market. Unlike the Australian export destinations that are low-cost Asian countries, the primary destinations of American coal products are European countries and some countries in America, such as Canada and Brazil (EIA, 2009b). As a result, although there is also increasing consolidation in the U.S., a lot of small operators still exist.[3]

The existence of so many small operators is a major barrier for the American coal mining industry in embracing the risk management model. On the one hand, the model itself does not fit well with the needs of small operators. It is unlikely that a small mine with only several employees will develop a highly formal and structured risk management system unless the law requires it to do so. On the other hand, the diversity of the industry results in a "different level of commitment."[4] Some companies "walk the walk," whereas others "talk the talk." It is very difficult to bring everyone together to drive up safety standards.

According to some informants, it is difficult to use some mechanisms to separate the good from the bad performers for two reasons. First, it is difficult to exert pressures on those mines located in remote places that do not care what happens at the national level. The industry reputation does not matter to them because they are not listed on any stock market. Normative pressure does not work well either. According to sociological theory, norms can be effective when people or organizations are bonded in a small community. However, some coal operators never participate in any association or forum. As an informant notes:

> It is hard to get all the mines together because there are so many. There are some operations who don't want that policy. Some family operations don't want to be involved in these things. Maybe the numbers hurt us. In another industry, maybe it is easier to get your arms around and bring everyone together and say look, let's be proactive before the government comes to us. Let's raise our standards. Maybe the mining industry has done it somewhat, but not everybody is on board.

Second, mechanisms recognizing better performers may generate unexpected results. For example, an informant worries that the media may paint the industry unfairly by looking at only those bad apples. He said:

> If the press picks that up, they will paint the entire industry that way, and they will tie all together. They won't say it but there are a few companies that are trying to take the lead and make it better. But the industry as a whole is bad!

Third, a company that raises standards voluntarily may hurt itself. As a vice-president of safety in a company said:

> I will give you an example; there is a shortage of skilled workers in the industry. We had a strict drug testing policy; if we find two guys who don't pass their test, they are terminated. They may go down the road and work for this guy who has no testing policy at all. If he needs workers, he is not going to implement a policy. [5]

As discussed above, the existence of large numbers of small operators makes the diffusion of risk management more difficult. Even the largest coal producers, however, also have troubles in introducing risk management in a highly prescriptive regulatory environment in the U.S. The next part will address how the regulatory climate influences the modeling process.

REGULATORY CLIMATE

Unlike Australia, coal mining safety regulation in the U.S. relies heavily on the prescriptive approach. Under this approach, government regulations specify very detailed standards, and inspectors check to see if these detailed standards have been met. According to the Mine Act of 1977, inspectors are required to inspect all underground mines four times a year and surface mines twice annually. This type of highly prescriptive regulation prevents companies from developing a risk management model. On the one hand, compliance consumes the resources that companies could use to adopt the model; on the other hand, such regulation creates a compliance mentality that is not compatible with the idea of risk management, which expects operators to proactively identify, analyze, and control hazards. These two reasons are further explained below.

Consuming Resources

After the Sago accident in 2006, the government regulations in the U.S. were changed. They did not, however, move toward the direction of risk management. Rather, regulations became even more prescriptive. In response to the Sago disaster, some new legislation was enacted at both the federal and state levels.[6] Congress enacted the Mine Improvement and New Emergency Response Act (the Miner Act), which amended some requirements of the 1977 Mine Act. The Miner Act was signed by President George W. Bush and took effect on June 15, 2006. It requires underground coal mine operators to:

- Develop and adopt a written emergency response plan, which provides:
    - o Post-accident communications (a redundant means of communication with the surface for persons underground, such as secondary telephone or equivalent two-way communication)
    - o Post-accident tracking (the system must provide for above-ground personnel to determine the current, or immediate pre-accident, location of all underground personnel and will be functional, reliable, and calculated to remain serviceable in a post-accident setting)
    - o A wireless two-way communication system and electronic tracking system within three years (by June 15, 2009)
    - o Emergency supplies of breathable air and caches of self-rescuers
- Make available two experienced rescue teams that are able to respond in one hour (NMA, 2009b)

After the Miner Act took effect, both MSHA and operators were overwhelmed in implementing the new requirements. In particular, some technologies required by the Miner Act are not available, e.g., the true through-earth wireless communication technology. Therefore, the experts in MSHA, NIOSH, and operators struggled to identify possible technology to meet the requirements by June 15, 2009. In the conferences and workshops in which I participated during the second half of 2008, many presentations were concerned with how to develop those new technologies.[7]

There is no doubt that the new regulatory requirements, particularly those for new tracking and communication technologies, will have a significant positive impact on safety when they are implemented. However, the new legislation and regulations at least postponed the industry's modeling of risk management. The reason is straightforward. The new legal requirements took away operators' time and resources. Taking Whitesburg Coal as example, as the last chapter described, it decided to develop an advanced risk management system at the end of 2005. Unfortunately, the Sago disaster occurred on January 2, 2006. According to an informant, the company's energy and time have been absorbed by dealing with government regulatory issues after the Miner Act was enacted. It has had to make the regulatory issues a priority because there is a rigid deadline. "Had all those government things not happened, we may have actually been there now," the interviewee said. It means that Whitesburg Coal's plan of developing a risk management system has been delayed at least two years. Its plan may be further delayed if the S-Miner Act is passed.[8]

In a broader view of the issue, the new legislation and regulations have consumed almost all of the operators' time and resources, which could be used not only to develop risk management systems but also to proactively conduct accident prevention (risk management is a strategy of accident prevention). According to some informants, a problem of the Miner Act is that all of its requirements are concerned with response after accidents occur; nothing in the Act deals with proactive accident prevention. For example, a respondent said:

> One point I like to make is just a personal view; when the Miner Act was introduced, it was very demanding, involving a lot of changes on the self-rescuer, emergency, rescue chambers, and providing additional air. The whole industry is focusing on the Miner Act. There are few things in the Miner Act to prevent accidents. The act deals with events when things happen. I fear, as I said early this week, the industry will take their eye off the ball—accidents, because everybody is consumed by this Miner Act.

Another interviewee also said that since 2006 safety professionals have spent about 90% of their time dealing with regulatory matters and only

10% on accident prevention. When talking about the Bluefield Coal Symposium he attended in August 2008, the interviewee said that it was the first seminar discussing accident prevention in the past two years. All of the other seminars were about complying with government regulations. He missed very much the time when they discussed accident prevention with MSHA before 2006.

> Prior to 2006 we were sitting in NMA, I can even tell you where I sat at the table. The administrator of MSHA was Ray McKinney at that time. We were talking about SLAM and behavior-based safety. He had a couple of MSHA officials with him. We had maybe 12–15 coal operator safety people. We had a group of maybe 20 people total. We were actually discussing back and forth what to do about training. That was a good discussion. I walked out of there saying "we are ready to change, and the industry is ready to make the change." Unfortunately, after Sago, that change happened but it went 180 degrees, the opposite of where it should go. To me, that is when you get to zero disaster. At some point you can go back to those kinds of conversations. Whether it will ever get back to them in my time, frankly, I am pessimistic. I don't think it will happen in my career, but at some point, it has to happen.[9]

Some readers may say that the influence of the new requirements will fade away after several years and then the industry can be back on track to develop risk management systems and conduct accident prevention. Yes, this could be. However, even dealing with regular inspections also consumes a lot of time and resources for safety professionals. During an interview, a vice-president of safety showed a daily report from the operations of his company.

> I got a list of a reminder of violations and a list of who the inspectors at the mine were. At the site in Cariky, Virginia, we had 12 inspectors on that day. When I look at the sheet, we had 21 inspectors in Kentucky and Virginia mines. So I know we had 21 inspectors and we got 10 citations yesterday. I got this everyday and we have inspectors at our operations almost everyday.[10]

When talking about the influence of MSHA on the effort to embrace risk management, the Vice-President of Safety in another company said:

> They are worthless, absolutely worthless. They have no idea what to do about accident prevention. All they care about is enforcement. In fact, they are counterproductive because they are taking resources away from the things we know would be helpful to reducing accidents. We are so compliance-based and we just don't have time to do the things we like to do.[11]

All of the interviewees mentioned above are from the largest companies. If they have limited extra resources available to conduct accident preventions and embrace the risk management model in such a prescriptive environment, smaller operators are likely to have even fewer.

Creating Compliance Mentality

More important, this type of prescriptive environment has created a compliance mentality in operators. This mentality is incompatible with risk management because people who only follow rules do not think. By contrast, risk management expects people to think proactively about potential hazards, and their causes, consequences, and control measures.

According to some informants, many people in the industry believe that "compliance drives safety."[12] They think that they have already dealt with risk by complying with statutes and regulations because these laws have identified the risks. This mentality is problematic because many accidents have demonstrated that laws cannot identify all the risks.

When talking about industry's compliance mentality, an interviewee said:

> They manage compliance and safety is somewhere else. All their focus is on the compliance side and on the enforcement side. They have no time and no resources to deal with the effective safety process. Compliance is a component of safety; we should not have compliance drives safety; we should have safety drives compliance.

He also gave an example of the kind of inspector behavior that can promote a culture of accident prevention that is compatible with risk management as opposed to a compliance mentality.

> If an inspector walks into my office or walks into a mine office, and looks at a power cable, specifically a 110-volt power cable that runs a refrigerator, and he sees that power cable has exposed conductors; that insulation is not there. He says "you need to fix that; that is a violation." Guess what, they fix it, but they never understand the implications of those exposed cables. This is part of the risk-based approach, as I told you. The other side of the inspector option; he says "Mr. Supervisor or Mr. Miner, do you see that cable right there that has exposed conductors going to it." They say "yeah." He says "do you recognize the hazards." They say "yeah, if I touch it, I will get electrocuted." "This is exactly right; realizing that kind of condition is good, it could hurt you or somebody else. That loss goes to the family and the company. We have to fix that. How do we make sure that doesn't happen again?" If you depend on that approach from the risk management perspective, as opposed to a compliance perspective, then you have a safety driving

> component; not only you fix the hazards and violation. The level of importance is much greater than it is if I tell them that it is a violation of law, and you have got five minutes to fix it.[13]

The discussions above explained how two structural variables—the characteristics of the industry and the regulatory environment—prevent the industry from embracing the risk management model. Beyond the expectations that I had before actual field work started, a modeling variable emerged based on my observations of the industry events that I participated in. I found that behavior-based safety (BBS) is more attractive to many American coal mining companies than risk management. The next section will explain how BBS, playing as a role of a competing model, inhibits the diffusion of risk management.

COMPETING MODEL—BEHAVIOR-BASED SAFETY

According to social cognitive theory, whether a model can be successfully adopted depends on a number of factors. Some are related to environment constraints as shown in the previous two sections. Others are associated with the attributes of observers. Still others pertain to the properties of the model itself.

Attributes of Observers

Observers' cognitive capabilities can limit the extent to which modeled information is perceived. In other words, the value of a model cannot be fully understood if it is too complex and beyond observers' cognitive capacities based on their previous knowledge. As the last chapter described, many people in the American coal mining industry do not understand the different levels of risk management, ranging from systematic advanced risk management to basic formal risk management (e.g., JSA), and to informal risk management focusing on employees' behavior. The coal mining industry in the U.S. is more familiar with the latter two levels than with advanced risk management.

According to Bandura (1986, 53), observers' perceptions are "guided by preconceptions." Their expectations, based on existing cognitive structure, can direct their attention to some things but not others. The expectations can also affect what observers "extract from observations and how they interpret what they see and hear" (*ibid.*). For example, many people in the U.S. coal mining industry tend to see risk management as an informal and behavior-oriented approach, which is at the lower level of risk management. They also say that they are doing risk management, but the risks they focus on are behavior-oriented and inherent in individual jobs, but not system-wide in nature.

Moreover, when some companies and MSHA are exposed to the Australian model of comprehensive risk management, they tend to simply take the informal parts but not the highly structured and systematic part of risk management. An example is the SLAM program, which has been adopted not only by many companies, but also by the federal government regulator, MSHA. Around 2004, the National Mine Health and Safety Academy introduced the SLAM process as an education and training initiative to both inspectors and the industry. Initially, the SLAM process that was developed in Australia refers to Stop-Look-Assess-Manage. It requires all employees to *assess* the risks and rate them according to a risk matrix before conducting any task, and then to *manage* all risks to as low as reasonably practicable (ALARP). According to some informants, SLAM was introduced in the U.S. because MSHA was increasingly focusing on the human factors on safety performance. MSHA employees, who had contacts in Australia or with Australian operations and venders in the U.S., brought this process to the MSHA.[14] What is interesting is that MSHA did not adopt the whole structure of the advanced risk management model used in Australia, but only a small part of it—SLAM. Even this program is actually a simplified version of the Australian SLAM program, although both share the same basic process. As an interviewee stated:

> At that time we were pretty deeply involved in the behavior-based program. We were looking into all the different areas. It [SLAM] was not exactly what they had in Australia, but they had the same concept and same approach. They [MSHA] took the abbreviation and ran with it.[15]

MSHA's training manual on SLAM provides evidence that the agency only extracted the simplest part of the Australian model of risk management. This manual says that "risk assessment is a simple process, with a big effect, to identify and control hazards" (MSHA, 2007: ii). Actually, the risk management model in Australia is a very complex process. In addition, based on the information in the MSHA's SLAM training materials, the risk ranking technique is very basic and does not involve a complicated risk matrix such as that used in the Australian version of SLAM (MSHA, 2008b).

As a result of this existing cognitive pattern in the U.S., which tends to accept informal and behavior-based techniques, many American coal mining operators, when they have built up commitment to higher safety standards and when they start to consider adopting any new model to improve their safety, immediately turn to a competing model of risk management—BBS. In the industrial events I participated in, the Bluefield Coal Symposium and the education sessions in Minexpo, some operators presented their BBS programs. However, no one from American companies talked about the model of risk management.

Therefore, it is clear that a competing model, BBS, in the American coal mining industry at least to some extent inhibits the risk management model from diffusing in the U.S. The reason is simple. If BBS is more salient and attractive to operators, they will pay less attention to the risk management model and also tend to interpret this model through a BBS perspective. The next sub-sections will explain what the BBS model is and what properties of this model make it more easily accepted by the American coal mining industry.

Behavior-Based Safety Model

Risk management and BBS as two competing models can be linked to the traditional "safe place" and "safe people" debates over how to maintain workplace safety. The former emphasizes engineering controls in unsafe working environments, whereas the latter stresses control of unsafe human behaviors. As shown in Chapter 2, the risk management model involves a hierarchy of control. It assumes that engineering controls such as installing physical barriers are more effective than modifying procedure and control of employees' behaviors in reducing hazards. Therefore, risk management to a large extent follows the tradition of the "safe place" logic. It does not mean that risk management ignores the importance of behavior control. It just emphasizes that operators should prioritize engineering control over behavior approaches.

BBS, on the contrary, follows the tradition of "safe people" based on the assumption that the majority of injuries and illnesses are the result of unsafe acts.[16] The contemporary behavior-based approach to safety is founded on behavioral science, particularly applied behavioral analysis emerging from B. F. Skinner's research in the 1960s (Geller, 1997). According to Krause (2001, 28), applied behavioral analysis was adopted in studying safety performance in the 1970s by some scholars and consultants. During that time, "behavior-based safety" was first termed by Gene Earnest and Jim Palmer.

As Krause (2001) suggests, the development of BBS has gone through three phases. Before 1985, almost totally relying on behavior analysis, BBS was top-down and management-driven. The primary focus was to change individual behavior through supervisors' audits and reinforcements. Between 1986 and 1996, the model changed dramatically because of the influence of the Total Quality Management (TQM) movement. As a result, BBS started to incorporate "TQM and organization development principles" into applied behavior analysis (p. 29). BBS became employee-driven and started involving behavioral observation and "feedback as an improvement mechanism" (p. 29). After 1997, in response to criticisms that BBS overemphasized worker behavior based on the long pervasive worker-facility dichotomy, BBS started engaging all employees, including managers, supervisors, and workers. The current model of BBS, integrating behavior analysis and knowledge of organization development, has become a comprehensive safety management system that involves not only behavior modification, but also system improvement and change of organizational culture.

Krause (2001) calls the current BBS model "second generation" BBS. Second generation BBS has at least the following characteristics. First, the model emphasizes employee-driven interpersonal observation and feedback. The purpose of observation and feedback is not only to change behavior, but also to cultivate a total safety culture (Geller, 1994). Second, the model focuses on positive consequences to motivate safety behavior, rather than on punishment of unsafe behavior. Therefore, peer observation and feedback include both unsafe and safe behavior. Third, second generation BBS must have continuous improvement mechanisms on systems using observational data. By analyzing such data, people can identify barriers concerning facility, equipment, design, and maintenance that prevent workers from working safely, and then they develop action plans to remove the barriers (Krause, 2001). In this way, second generation BBS has crossed "the bridge from a worker-focused strategy to a system-focused approach" (Krause, 2001, 30).

Dr. Thomas Boyce of the Center for Behavior Safety gave an example of how behavior observation can contribute to changes in equipment design. At a manufacturing facility, back strains had become common. Based on observations, people at the facility found that most problems were caused by lifting boxes onto a conveyer for loading onto trucks. Workers usually twisted their torsos to move the boxes onto the conveyor, but they did not change their feet positions. Back injuries were the result when workers performed the task this way for ten hours. One member of the group found a simple solution by redesigning the station. He designed a $100 fix-a-line attachment that transports the boxes to the conveyor, so that workers do not need to twist their torsos.[17]

This "back end" of BBS using observational data to develop action plans for improvement is the most important advantage of BBS. This is because design teams cannot predict every potential problem that arises in the actual use of equipment. BBS allows people to continuously assess and improve design and equipment on the shop floor (Krause, 1997; 2001). As Krause (1997, 22) notes, "anything less is not BBS."

In sum, a typical BBS process involves the following steps: (1) identifying and defining critical safety-related behaviors based on reports, near-misses, and experiences; (2) observing and measuring the frequency of those behaviors through a pilot group or a larger group of employees; (3) providing feedback and setting improvement targets; (4) developing action plans for continuous improvement using observation data.

From this description above, it can be seen that the distinction between "safe place" and "safe people" is a false dichotomy in theory. This is because, if used correctly, both the risk management model and BBS deal with systems as well as workers' behaviors. The difference is that risk management starts by identifying and analyzing risks in systems and breaks them down to workers' behavior on the shop floor level, whereas BBS starts by observing workers' behavior and moves up to the barriers in systems that prevent workers from performing safely. However, this dichotomy can be real in practice. On the one hand, the dichotomy of "safe place" versus "safe people" is to some extent the result of political rhetoric. It is all about whom one blames. Often, employers advocate "safe people," whereas laborers emphasize the importance of engineering controls. The information I collected suggests that the tendency to transfer blame is real. For example, an informant from a union said that the union welcomes employers adopting risk management to assess and control physical risks, but not behavioral risks. By contrast, an interviewee at the management level gave the following comments when talking about BBS:

> MSHA looks at it as a process that Company A [the name is fictitious] is responsible for mine safety. I can mess up, I can get myself killed, get myself hurt, but whenever it comes down to who is at fault. Company A is at fault. But in a BBS process, if I get hurt or I get killed, 96% of time it is because I have done something wrong. My company does not expect me to do that. Other companies are the same way. They don't expect people to do that.

At the same time, even without the tendency to shift blame, priority often has to be placed on either people or facilities, given resource constraints.

The two models—risk management and BBS—have both advantages and disadvantages in terms of the impact on safety performance if used alone. Compared with the model of risk management, BBS has several advantages. First, BBS has the ability to reach day-to-day workers' performance on the shop floor. Because of the emphasis on documentation and hierarchical control, the risk management model may result in disconnect between what is on the paper and workers' actual performance. Evidence can be found in both government documents and interview data. For example, The NSW *Wran Mine Safety Review* identifies an "apparent 'disconnect' or 'disjunction' between corporate management systems/plans and their translation to a working level at the 'coal face'" (Wran and McClelland, 2005b, 49). Similarly, as a vice-president of safety in an American coal mining company said, when he visited a coal mining company in Australia, the safety staff were keen to show him how comprehensively the risk assessment had been done. But when he went underground he saw hazards that would never be tolerated in his operation in the U.S.[18]

Related to the last point, the second advantage of BBS is associated with its employee-driven characteristic and its emphasis on feedback. Risk management systems, as a key component of systematic OHS management systems, are hierarchical by their nature. This is partly because OHS risk management systems have their origins in top-down management approaches, such as quality control management (Frick et al., 2000).

Because of its top-down character, the risk management model may result in an "authoritarian approach to management," which acts as "a barrier to genuine worker involvement" (Gunningham and Johnstone, 2000, 145; Saksvik and Quinlan, 2003, 44). This can be demonstrated by evidence in interview data, government documents, and other scholarly literature. For instance, the interviewee mentioned above said that people who conduct risk assessment in Australian mines were always those above the manager level based on his observation; they never communicated to the miners. Moreover, according to the NSW *Wran Mine Safety Review*, unions claim "a lack of effective direct worker and union consultation" in "OHS management systems, including risk assessment" (Wran and McClelland, 2005b, 55). Likewise, Gunningham (2008, 351) finds a big gap between workers' statutory rights being consulted and actual involvement in risk management in the Australian mining sector. Third, BBS is more effective in handling minor injuries and illnesses, such as back and knee injuries, because of its ability to reach the behavior level compared with risk management, which emphasizes engineering controls.

In contrast, the major weaknesses of BBS compared with risk management are twofold. First, BBS may distract attention from facilities, maintenance, design, and engineering improvements if managed inappropriately (Krause, 1997). An informant pointed this out when talking about some possible by-products of moving toward zero harm.

> A danger for the risk management idea, in my view only, is that risk management would get wiped out in the attempt to achieve zero harm on day-to-day injuries. . . . My view is when you try to achieve zero harm, you are looking at day-to-day activities; where risk management is looking at a broader, bigger and longer range of views. . . . There are people putting together a program called SLAM. Essentially, that is just hazard analysis, hazard identification. They were calling it risk management, some people. But I think that is not really true. To get to zero harm, a lot of your efforts are focused on hazard identification, a different level of concern. In my view, risk management has a higher elevation level. That is hopefully how to put our program together. Our final program will separate that way.[19]

Second, although BBS is good at dealing with minor injuries and illnesses, it has limited value in preventing major hazards such as roof falls, fire, explosion, and inundation. As an informant described, BBS can help people to be aware of hazards. However, "proper body position" and "personal protective equipment" cannot "stop 500 tons of rocks falling on people's heads." The major hazards like this should be dealt with through a "combination of engineering mining methods and organizational decision making."[20]

The two models actually can be used to complement each other. Advanced risk management is driven from the top and focuses on major hazards, whereas BBS pushes from the bottom by dealing with dramatic and minor injuries and illness. As shown in the last chapter, developing both models at the same time to deal with different levels of issues is exactly what Whitesburg Coal has decided to do.

Nevertheless, for the purposes of this study, it is less important to understand the difference between these two models in terms of their functional value for improving safety, rather than to identify what properties of the BBS model have led to its higher popularity in the U.S. The next sub-section takes up this question.

Properties of the Competing Models

As previously discussed, the properties of a model itself can influence the extent and speed that observers adopt the model. The following properties of BBS evidently have made it easier to be adopted by the American coal mining industry than the model of risk management.

Popularity

The behavior-based approach is more popular in the U.S. than in other countries for at least two reasons. First, the behavior-based approach to safety has a longer intellectual tradition in the U.S. According to Frick and Wren (2000, 21), it can be traced back to the Safety Movement, which applied Taylor's scientific management idea to accident prevention in the beginning of the 20th century. The Safety Movement "further developed into the Loss Control approach of Heinrich (1931)." The Safety Movement emphasized that accident prevention "should both motivate and control workers to follow the detailed work procedures prescribed by managers and safety professionals" (*ibid.*). Second, the greater popularity of BBS is partly because union representation in the American workforce is relatively low. According to Bean (1989, 157–180), the average union membership representation in the U.S. was about 24.6% of total labor force in 1970–1985, which was one of the lowest in the Organization of Economic Cooperation and Development (OECD) countries. During the same period, unions represented more than 50% of employees on average in many other OECD countries, such as Australia, Belgium, Denmark, Norway, and Spain. In 2007, trade union density was only 11.6% in the U.S., ranked second from the bottom in 16 OECD countries (Figure 4.3). This lower level of union representation in the U.S. gives employers more room to implement BBS because, as previously stated, unions are the major force opposing BBS. Unions tend to see it as a tactic that employers use to transfer blame to workers.

Because of the longer intellectual history and less union resistance, U.S. corporations have placed a stronger emphasis on workers' behavior than have those in other countries. For instance, Frick and Wren (2000, 27) note that "control of workers' 'safe behavior' remains a significant aspect

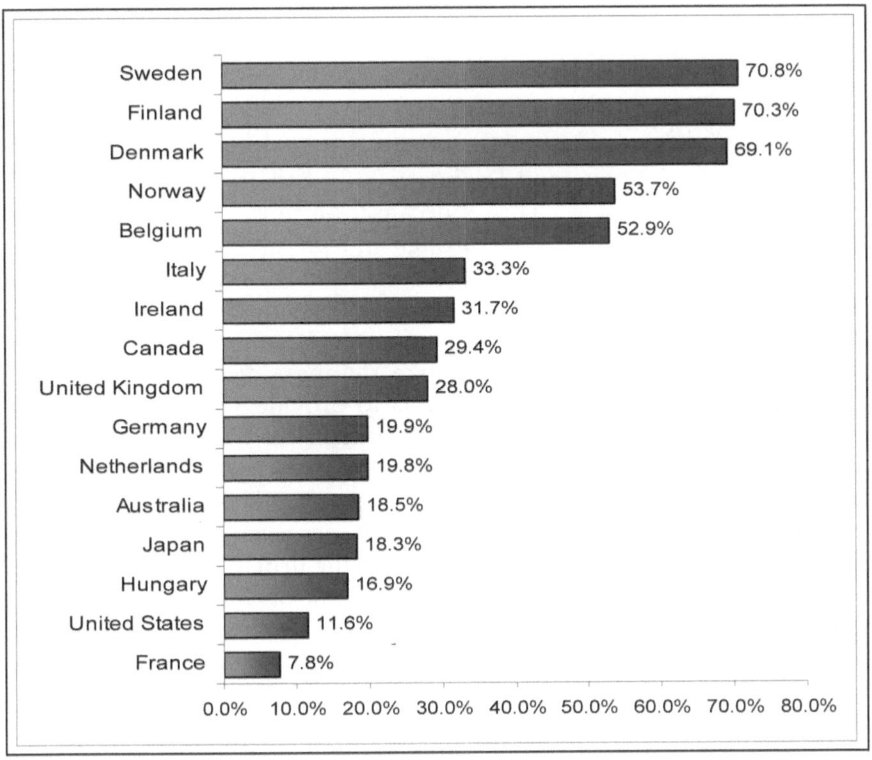

Figure 4.3 Trade union density in OECD countries in 2007.
Source: OECD (2009).

of many OHSM systems that are marketed, particularly those developed by U.S. corporations." The best example is the DuPont safety management system, which "aims to modify workers' behavior, rather than avoiding hazards through engineering" (Frick and Wren, 2000, 28). These scholars discussed the first generation of BBS. The current DuPont safety management system has evolved considerably. However, this information at least shows that the behavior-based approach has been very popular in the U.S.

Other evidence of the popularity of BBS in the U.S. is the development of Behavioral Science Technology, Inc. (BST), which is a California-based company founded in 1979 by Thomas Krause and John Hidley. BST is perhaps the most recognized name in the field of BBS. Quoting Krause (1997, 21), the founder of BST, "BBS has become quite popular in the U.S. and Canada, and awareness of it has continued to grow throughout the rest of the Americas and in Europe and Australia."

As a result of the popularity of BBS in the U.S., people in the coal mining industry are much more likely to be exposed to this model and to acquire related knowledge. The later section will provide an example of how BBS was introduced in one coal mining company, Blacksburg Coal.

Depth of Expertise

According to Braithwaite and Drahos (2000), the depth of expertise is a critical factor in diffusion. This is because, on most occasions, diffusion involves observational learning through modeling. To adopt a model in practice, adopters have to go through the process of acquiring knowledge, processing information, transforming it into symbolic form, and implementing the model in practice. Adopters can certainly acquire and process the information about a model through the mass media. However, they usually need help from external experts to provide knowledge and techniques on how to apply the new model in practice. Therefore, the greater the expertise about a model in one place, the greater the chance the model will be adopted.

Many BBS model missionaries and mercenaries, such as scholars and consultants, exist in the U.S. One famous consulting firm is BST mentioned earlier. Another well-known scholar and consultant on BBS is Scott Geller at Virginia Tech. He has assisted at work sites across a wide range of industries in developing BBS programs. During interviews, many informants from the coal mining industry mentioned Geller when they talked about BBS. During the time of my research, some major coal mining companies chose different consultants to work on their BBS programs. For example, BST helped Arch Coal, the second largest coal producer in the U.S. They gave a joint presentation about their BBS program at the Minexpo in Las Vegas in September 2008.

Unlike the BBS model, the depth of expertise on the risk management model is rather thin in the U.S. As a vice-president of safety in a coal mining company said in an interview, the difficulty in finding a consultant is a barrier in developing a risk management system.

> Finding a company in the U.S. that can actually come in and do the kind of training we look forward to is very difficult. Australians are expensive to bring here because they have to travel back and forth. We have to work on finding somebody that can teach and train on how to do risk assessments.

This problem is also mentioned by a risk management scholar and consultant in Australia.

> If I worked as a risk management consultant, I would move to the U.S. The U.S. is ripe for risk assessment in mining. . . . There are very few people in the U.S. who understand risk management and mining. Many companies look for expertise and find considerable risk management expertise in oil, chemicals, space, and the military but they are

not familiar with mining. That may be one of the reasons why progress is slow." [21]

The existence of thick expertise on BBS and thin expertise on risk management is the environment that coal mining companies encounter when they seek to adopt new models to improve safety performance. However, this environment is not a given but has formed based on people's repeated choices. According to Bandura's (1986) social cognitive theory, people's psychology, behavior, and environment are interdependent. An example is that the actual televised environment depends on what people choose to watch. People's viewing behavior based on their preferences shapes what television stations offer because of commercial considerations. The televised environment, in turn, partly shapes people's preferences (p. 24). As with this example, the U.S. environment with plenty of expertise on BBS is also a result of people's consistent choices to look at behavioral issues based on their cognition. The more such choices, the more model missionaries and mercenaries will provide services on BBS.

Complexity

According to Rogers (2003), the perceived complexity of an innovation influences its adoption rate. Whether a model is perceived to be complex or not largely depends on people's cognitive capacity. To most people in the American coal mining industry, BBS is relatively easier to understand than risk management partly because BBS does not involve complicated risk management techniques, such as bow-tie analysis, and partly because behavior-based approaches have been used in the industry for a long time, such as JSA.

Although it also requires advanced knowledge in behavioral science to develop an effective BBS program, BBS demands less expertise and knowledge than risk management to start with. As Bandura (1986, 51) indicates, "amenability of an innovation to a brief trial" is a key factor in determining the rate of adoption. In other words, a model that can be tried on a "limited basis" diffuses faster than one that has to be tried on "a large scale with substantial efforts and costs." Since risk management is a highly structured and comprehensive approach, operators have to commit a large amount of time and people to conduct risk assessment. During an interview, an informant used an analogy of quitting smoking to explain the difficulty of adopting risk management at the early stage.

> If you do smoke, I can sit here and tell you that you will feel a lot better if you stop smoking. But really you don't for a long time. It is very hard work. It is the same sort of thing. I can say risk management is good for your business. But the initial phase of planning is actually hard work.

By contrast, with BBS, employees can start to do behavior observations after a few days of training. That is at least part of the reason why BBS is more likely to be adopted in small operations than the risk management model.

Rewards

As Bandura (1986, 162) argues, a model is more likely to be adopted when it can "produce prompt and observable benefits, and when the causal relationship between new practices and results can be easily verified." Usually, results can hardly be seen until a new practice has been used for a while. However, the more delayed the observable benefits, the less motivation the adopter has.

Compared with the risk management model, BBS can produce more prompt and more observable benefits, and the causal relationship between BBS and outcomes is easier to verify. This is because the two models deal with different levels of risk. As described earlier, BBS deals primarily with employees' day-to-day performance of their jobs, whereas risk management mainly targets major hazards that can cause multiple fatalities. It is more common to see accidents that result in injuries, such as cut fingers and twisted knees than disasters such as fires, explosions, and roof falls. In an operation, these disasters may not occur in many years. Therefore, it is difficult to verify the causal relationship between the adoption of risk management and better safety performance because many factors may change during such a long period, such as improvement of technology.[22] With BBS, adopters can see the improvement of performance in a relatively shorter time. This difference between the two models was highlighted by many informants. For example, every interviewee in Blacksburg Coal (whose process of adopting BBS will be discussed below) repeatedly mentioned the motivation resulting from the observable inverse relationship between the number of observations and the accident rate. In other words, "as the number of observations goes up, the accident rate goes down."[23] One informant called this observable relationship the "fuel" of their BBS system.

By contrast, another interviewee noted the difficulty of verifying causal relationships between risk management and results.

> Even if you decrease the risks, you see the decrease of indicators like injuries and fatalities, but you may not relate that to risk management. If you only measure absence or something, it is not like production going on, it is hard to measure success—wow, we should do it again because production is going up, see . . . it is better. With risk management it is hard to tell.

Compatibility

According to Rogers (2003), perceived compatibility is another determinant of the diffusion rate of an innovation. Compatibility issues exist at the levels of both organization and the broader society. Here, the focus of interest is the compatibility of the two models with American culture. Compared with BBS, the risk management model is less compatible because of at least the following two aspects. First, the risk management model's strong emphasis on formality and documentation does not fit well with the individual-oriented U.S. culture. Some informants expressed that the Australian model of risk management is too formal, and some informal risk management styles, such as BBS, fit better with American culture, because it emphasizes that individuals can make a difference. An interviewee who emigrated from England to the U.S. provided the evidence of this distinct American culture based on his observation.

> When I came to the U.S. from Britain 40 years ago, I noticed America is far more informal compared with Britain or Europe or Australia. It was the same with quality. ISO quality standards came out and were used extensively in Europe and took a longer time for the formality of that to come to America. The same is true with safety. I think Americans are very safety conscious, but Americans like to do things informally. Perhaps one of the reasons for that is individuals can use their own talent to put in their systems. I think the American view is we want to have safety and production, we want to use the best technology. But the way of doing it by filling out a lot of paper work and going through a formal procedure is not the best effective way to achieve the result.

Different informants expressed similar opinions. For instance, one vice-president of safety pointed out that Americans tend to be less willing to follow a standard structure.

> I think that is part of the difference between American culture and Australian culture. Australians are more willing to follow . . . if you have a standard out there, they are more willing to follow the standard. Americans are more free thinking. I think behavior-based safety fits our culture a little bit better because when you talk to people in behavior-based, you have to go down to what is really meaningful to people, what really motivates them, how they think, why they act the way they do. You have to do that in the U.S. In Australia you pretty much tell them what to do, here is the standard and do it and they do it . . . the culture [of America] is I know what I am doing and leave me alone. Our people are not team based yet, not collaborative, they are all just kind of like cowboys. . . . I know what I am doing, don't tell me.[24]

An interviewee from Whitesburg Coal indicated that he saw this kind of "push back" when the company tried to introduce structured and documented risk management systems.

Second, the risk management model's strong emphasis on formality and documentation is not compatible with the litigious culture in the U.S. This is because a typical risk assessment produces a documented risk matrix in which risks are ranked according to their likelihood and consequences. More attention will be paid to higher ranked risks. However, if a lower ranked hazard occurs, someone may hold the company liable because the document is evidence that the company had been aware of the risk, but had given it inadequate attention. This is why, even in Australia, some operators favor a prescriptive approach because it is not their problem when something happens as long as they comply with the law. Therefore, many American operators prefer informal risk assessment without documentation to avoid legal liability.

Some respondents raised this problem. For instance, one interviewee considered the fear of liability as one of the major hurdles to American operators' introducing risk management.

> One [hurdle] would be to get past the view that if I do this I may be setting myself up for second-guessing if I am wrong on my assessment. There will be reluctance by some people. . . . No, I don't want to do that, as soon as you talk about formality, it is going to be a written and formal process and people start saying what will happen if I make a mistake. I give you one example that came out in our first discussion. We are going to have a list, right? Yeah, you look at consequence and probability, rank and prioritize them and work on first one first and last one last. That is standard risk assessment. A guy says "what happens if I work on first and second, and the seventh happens. We listed it as an issue." That is the hurdle. If I say there is a problem that should be looked at, I haven't gone around to it yet, I can be in trouble. We are a litigious society. When you do it ad hoc, there is no threat.[25]

This fear of legal liability because of the documented risk assessment can also be demonstrated by the litigation against OSHA's national Cooperative Compliance Program (CCP) initiative. The program was launched in 1997. According to Needleman (2000, 79), OSHA identified around 12, 500 employers with unusually high lost workday injury and illness rates based on survey data. OSHA put these employers on its primary inspection list, while providing most of them with an alternative. If they chose to join the CCP and voluntarily developed safety and health plans, they would be removed from the primary inspection list. Some employer associations, such as the U.S. Chamber of Commerce and the National Association of Manufactures, challenged CCP in the U.S. Court of Appeals for the District of Columbia Circuit in 1998. In April 1999, the court struck down the CCP on the grounds that it "amounted to a new standard, forcing employers to do more than what is legally required by the OSHA Act and therefore should have been subjected to public notice and comment" (Needleman, 2000, 80). However, according to Needleman (2000, 81), the liability concern was "central in trade associations' litigation against" the CCP. This is because the CCP required employers to identify hazards in their workplaces and develop a plan to control those hazards. This evidence removes "an existing legal defense for employers facing OSHA citations" by arguing that hazards are not recognized (*ibid.*).[26]

Also, related to the liability issue, risk management can put management in a dilemma. Suppose the management receives a list of recommendations according to risk assessment; if they do not act on them, they may have trouble if something happens. However, if they implement them, they may be blamed for spending too much money if nothing happens. An informant gave an example.

> Risk says we need to change what you are doing here. We need to add thousands of dollars worth of equipment; we need to do this, this, and this. We think this is the highest priority in your coal mine. You sit there and say I don't have budget for that. That is the hard part. We have seen that. We have a property insurance group doing some risk assessment. They tell us this is what I want you to do, we think you need to do a b c d e, the operation says I don't want to do it. It does not seem a high risk to me. It is a super expensive project. I don't want to do it. Now if something goes wrong here, they will all sit there and say these guys are sending me out; if I don't do it and something happens, I will get all these memos, that is the fear.

Because of the reasons discussed above, including different properties of the models and attributes of people's cognition, BBS has been more easily accepted by the American coal mining industry. Whitesburg Coal's experience supports this observation. According to informants from this company, when they tried to present both models to the affiliated operations, the program involving BBS was "localized" more than risk management. This is because people in the operations are more familiar and comfortable with BBS. When they use behavior-based techniques, "they walk in their own shoes." By contrast, people in the company are not familiar with risk management and not sure how it fits into their existing practices. They also see pushing back because of the fear of legal liability.

The relative advantage of BBS compared to the risk management model does not mean there is no resistance at all to BBS in the U.S. As a matter of fact, unions, particularly the United Mine Workers of America (UMWA), are strongly against BBS because they see it as "nothing more than a means to transfer blame from employer to the employee."[27] However, since the majority of coal mining operations in the U.S. are non-unionized—the operations in the two companies where the research was conducted are predominately union-free—BBS has plenty of room to be implemented in American mining companies.

Reflecting the issues discussed above, the next section will use Blacksburg Coal to illustrate how and why the BBS model has been chosen instead of the model of risk management. This case will be compared to Whitesburg Coal, which has introduced the risk management model.

Modeling BBS in Blacksburg Coal

Blacksburg Coal is one of the largest coal producers in the U.S. According to information on its website, it was founded in 2002 through the acquisition of coal operations owned by other companies. Currently, it owns 11 preparation plants and 65 active mines in the Appalachian region. Most of the mines are small or medium-sized.

How BBS Was Chosen?

According to informants, Blacksburg Coal's BBS program started around 2004 in some operations in Virginia and West Virginia. The Gerrard Mining Company in southern West Virginia,[28] where most field studies were conducted, is one of the earliest operations to introduce BBS in Blacksburg Coal.

A key driver for Gerrard Company to introduce BBS was two fatal accidents that occurred in March and April 2004. Every interviewee in this company mentioned the influence of these two fatalities on its BBS program. They referred to these two accidents as their "stepping stone" into their BBS process.[29] An interviewee told the story of the two fatalities.

> In January 2004, we started having the accidents, which carried on into February. In February we already kind of surpassed the number of accidents we had in the preceding year. We knew there was something that was going wrong, but we just couldn't quite put a finger on it and thought it was more or less just fate. Then in March of 2004, we lost a very dear and loved employee at the preparation plant. To this day—that was four years ago—if you bring out this guy's name in the preparation plant, you are going to see some eyes start to tear up. He is very well loved. We lost him and a month later we lost another employee. Between the two fatalities, we were kind of looking at if you mine coal these things happen, which goes back to the mindset 20 years ago. If you mine coal, an accident is going to happen. When we had the second fatality, a lot of us started to think about things that we could do to improve safety.

As the quotation above illustrates, this is the process of change in mentality described in the last chapter. When talking about the significant impact of the fatalities on their mentality, another informant said:

> That changed everything toward thought of safety. You might take your guard off when you work on some. If you are in a hurry, you lay

part of the guard off and you put it down there. All that stuff ceased. It just opens your eyes. This can really happen.

This case again demonstrates that the process of driving up standards in companies is not only proactive but often reactive in response to disasters. After building up the commitment to higher safety standards, Blacksburg Coal began to seriously consider how to improve safety performance. The question then becomes why they chose BBS but not the risk management model.

As shown previously, there are three steps in observational learning by modeling. First, observers have to be exposed to the model. Second, they transform information into symbolic form using their cognitive skills. Third, they implement the model into practice. Factors in each step can influence an observer to choose one model over the other. The following reasons may explain why Blacksburg Coal has chosen BBS rather than the risk management model.

First, Gerrard Company has been exposed to BBS but not to the model of risk management. According to informants, although the company did not bring in any external consultant, such as BST or Scott Geller, to help develop the BBS process, some key people at the management level had been exposed to BBS in other companies where they worked before they came to Blacksburg Coal. BBS has been applied in the U.S. for a long time, and many coal operations had some BBS programs in place by 2004. In addition, as an informant from Gerrard Company said, after the two fatalities, they invited some people from its sister company to explain the BBS process. This company had started its behavioral observation process before the two fatalities happened in Gerrard Company.[30] This reflects that, as mentioned earlier, modeling in one place can be a stimulator of modeling in other places.

By contrast, the management at Blacksburg Coal has limited exposure to the risk management model because they do not have strong interpersonal linkages with risk management model missionaries and mercenaries in comparison with the safety professionals at Whitesburg Coal. Blacksburg Coal's CEO stayed in Australia for one or two years. He may know about the Australian model of risk management. However, few safety professionals and other senior management people know much about the risk management model. The current Vice-President of Safety served as an MSHA representative in an international body. He has been exposed to the risk management model through his contact in Australia, but he had just been working at Blacksburg Coal for a few months at the time of the study. The other two interviewees at the senior management level at the Gerrard Company had never heard about the Australian model of risk management. Moreover, unlike Whitesburg Coal, Blacksburg Coal did not join Jim Joy's risk management workshop held by NIOSH. It also did not participate in NIOSH's MHRM project, which gave Whitesburg Coal the opportunity to have a big jump in developing its risk management systems.

The second reason is related to the management's cognition at Blacksburg Coal. As previously mentioned, many people in management had experience working with different safety programs, including BBS, in other companies before they joined Blacksburg Coal. According to informants, these people at the management level, particularly the Vice-President of Operations, saw BBS as being better than other programs because it can produce better outcomes. The Vice-President of Operations is one of the key advocates of BBS at Blacksburg Coal.

In addition, some of the management people at Gerrard Company do not share the perception of John at Whitesburg Coal, who sees risk management and BBS as tools dealing with different levels of issues. They tend to see risk management through BBS. One interviewee said:

> I don't know how to explain risk management other than through a behavior-based approach. Australians probably would not understand BBS unless they go through a risk management approach. . . . Risk management, BBS, and SLAM are all the same thing.

To some extent, his viewpoint is correct because any model has to depend on people's implementation to achieve its potential. He said:

> I think risk management and BBS have to start from here. It has to start from the CEO, all the way down to Jeff, and from Jeff all the way down to . . . or across different areas. Risk management and BBS start with the person. Not any level above or below.

However, viewing BBS as the same as risk management prevents Blacksburg Coal managers from recognizing the particular value of the risk management model at the systems level.

In addition, according to social cognitive theory, people learn not only through observational learning by modeling, but also through enactive learning based on their direct experience. The two fatalities provided Blacksburg Coal's officials with an opportunity for enactive learning. According to the MSHA disaster report, at least one of the fatal accidents clearly pointed to at-risk behavior. On the website of the MSHA library, a brief narrative about the second fatality that occurred in April 2004 recounts:

> A maintenance man was fatally injured at a sand and gravel operation. A three man crew was attempting to unplug material at the discharge chute below a 30-ton hopper. The victim entered the hopper from the top without wearing a secured safety harness and lanyard when the material suddenly gave way and engulfed him (MSHA, 2009b).

By contrast, as shown in the last chapter, John at Whitesburg Coal also had experience with enactive learning when he was involved in a belt fire disaster. However, his enactive learning pointed to the need of risk management. Therefore, at least to some extent, it is the different enactive learning experience in these two companies that led them in different directions.

Blacksburg Coal has not adopted the risk management model primarily because of lack of exposure and the way people perceive BBS and risk management. However, successful observational learning through modeling is not only about adoption. A failure of implementation that creates poor outcomes can lead to a search for alternative models. In contrast, if a model is successfully implemented and becomes ingrained in an organization, it can serve as an inhibitor of other models. The next section will show how BBS has been successfully implemented in Blacksburg Coal.

Implementing BBS in Blacksburg Coal

In this section, a brief introduction to the BBS process at Blacksburg Coal will first be given. Then, discussions will focus on some of the factors that have led to the successful implementation of BBS.

Blacksburg Coal's BBS process contains the common features of BBS, such as being bottom-up and employee-driven, having an emphasis on peer observation and feedback, and encouraging positive consequences. The process starts from behavior observations at the mine sites. All employees are encouraged but not mandated to carry a small observation card, write down any safe behaviors and at-risk behaviors they observe, and then turn the card in.[31]

To promote open communications and information sharing, the information collected based on observation is shared at different levels. At the mine site level, all of the observations, along with accidents and violations, are discussed in a monthly meeting in which mine managers and some employee volunteers participate. As a result of the discussions, they identify lessons learned, solutions, and suggested safety talks, which means an informal meeting between managers and employees to discuss certain safety items.

All the information that comes out from the monthly meeting at the mine site, including observations, accidents, violations, and lessons learned, is then brought to a monthly performance group (PG) meeting at the business unit level. The participants in the one-day-long PG meeting include the president and safety managers from the business unit office, as well as mine managers and representatives of employees from all the operations in the unit. The coordinator of the BBS process at each mine site gives presentations at the PG meeting to share information collected from operations with people from other operations. The coordinator is usually an employee who voluntarily takes the role. In the PG meeting that I attended, the presenters from mine sites included foremen, electricians, and shuttle car operators. The presentations usually follow the format of going through accident and violation review in the past month, lessons learned from the accidents and violations, number of observations in each category, lessons learned from at-risk behavior, safe behavior, recommended safety talks, new safety action plans to correct accidents, violations, and at-risk behaviors, and progress of previous safety action plans.

In addition, a safety task force conference call is conducted every two weeks across the company. The CEO, Vice-President, and General Counsel, joined by the presidents and safety managers from each business unit, chair these conference calls. Important issues faced by each business unit are discussed on these conference calls.

From the description above, it can be seen that BBS at Blacksburg Coal is not only a program for behavior modification. Rather, it also plays a role as a key node that links with other elements in safety management systems, particularly information sharing and lessons learned. Most solutions and lessons learned, which came out of accidents, violations, and at-risk behaviors, are related to behavior and procedures; yet engineering solutions such as improving equipment and design were also considered. This illustrates that the Blacksburg Coal BBS process is a second-generation BBS.

The success of BBS at Blacksburg Coal can be attributed to a number of influences, such as leadership commitment, engagement of workers, and adequate feedback to workers' observations. However, two other factors that are related to the modeling perspective are emphasized here.

The first is that BBS has produced observable benefits. Because of the observable benefits of BBS at Blacksburg Coal, an informant said that the BBS process can have a "lasting effect" and can "feed and fuel on itself." BBS has produced two major observable benefits in this company. The first is the causal relationship between the number of observations and a decrease in the accident rate. During the safety meetings I attended, presenters often used figures to show the inverse relationship between the number of observations and the accident rate in a certain period. The second is that BBS can contribute to lessons-learning based on observing data. An interviewee gave an example of how lessons learned by an operation can influence other operations through this information sharing process.

> Mine A [the mine's name is fictitious] had an accident sometime right after we started this process, where a guy got his finger amputated on an installing belt. We had the metacarpal glove policy in place, which may have prevented that accident or lessened the severity of the accident. But during the investigation, we found out the guy wasn't wearing his metacarpal gloves. Now the reason that he wasn't wearing metacarpal gloves is he got them saturated with water and mud and could not wear them anymore. The glove was outside and he was inside and could not get gloves. That was the root cause of the accident. What Mine A came out with on their EIG team was to take gloves inside and have gloves, safety glasses, and ear protection available for the labor inside. They would not have to go out. When we got to the performance group

> meeting, Mine A shared that information with their group and other groups have seen the necessity to do that and did it also. That is the way we set the process up. It is just fact finding and sharing the facts.

The second reason for successful implementation is the compatibility between BBS and Blacksburg Coal's culture. Based on the observations, what distinguishes Blacksburg Coal's culture from that of other large companies is its greater openness and informality. Through observation of the safety meetings, I found that the managers had lunch with workers and everybody could joke with one another. An informant said that "the people here are like a family and really care about each other." Related to this open culture is respect for employees. As a result of this open and respectful culture, employees are not reluctant to speak about their concerns. As an interviewee said:

> Employees have a say, just as anybody in this operation. The plant operator, electrician, load-up man—their say is really important. They bring something out, and a lot of action has been done to it. The people here all go to the same direction. They don't fight each other.

Openness and respect are the basis of the successful implementation of the BBS process. This is because all of the virtues of BBS rely on employees' participation. If their opinions are not respected by the management, they will not participate anymore.

In addition, unlike some large companies, Blacksburg Coal has a less formal management style. This is partly because most of its operations are small and medium-sized mines. It is not difficult to imagine that the management style of a company with 60 mines having dozens of employees in each would be less formal and structured than a company that owns ten mines, having hundreds of employees in each, like Whitesburg Coal. As Coleman (2008) shows, there are three Whitesburg Coal affiliated operations on the list of 64 major U.S. underground coal mines in terms of production. They are ranked in the top 40 (one in the top ten). By contrast, Blacksburg Coal has two operations on the list. Both are ranked lower than 45.

According to an informant, Whitesburg Coal's culture was already rigorous and paper- and documentation-oriented before it decided to introduce an advanced risk management model. Therefore, employees do not feel that it is a major change to apply the highly structured and documented advanced risk management model. Although Blacksburg Coal has not tried the risk management model, I expect that BBS fits better with its informal and open organizational culture.

Readers may ask "Has Blacksburg Coal's informal organizational culture led to the adoption of BBS or has the adoption of BBS changed the organizational culture?" According to informants, at least at Gerrard Company, the culture had been the same as today before it introduced BBS in 2004. It is evidently the culture that facilitated the adoption of BBS at Blacksburg Coal.

CONCLUSION

In sum, the research has found three barriers that keep the American coal mining industry from fully embracing the risk management model. Following structural explanations, two barriers are identified. First, the lower degree of international competition and relatively lesser importance to the nation's economy have resulted in a diverse industry structure, which makes the diffusion of risk management more difficult. For the same reasons, there has not been a strong feeling of common fate. Consequently, the industry cares less about its reputation and is less motivated to adopt the best practices. Second, highly prescriptive regulations create a mentality of compliance and consume the resources that companies could use to develop a risk management model. The third barrier is the existence of a competing model—BBS—that prevents the diffusion of risk management. BBS is more easily accepted by the American coal mining industry because of people's existing cognitive pattern that tends to look at risks at the micro level, such as workers' performance of each task. It is also because BBS has several favorable properties compared with risk management. BBS is more popular, has thicker depth of expertise, is more compatible with American culture, and produces more observable benefits.[32]

These findings have several implications for studies of the rise of regulatory capitalism. First, as some existing studies have suggested, regulatory capitalism, particularly industry's self-regulatory regimes, is more likely to grow in an environment in which mega-corporations are created and a strong feeling of common fate is recognized. Second, somewhat different from Braithwaite's (2008) argument that government regulation enabled the change toward regulatory capitalism, findings in this research illustrate that government regulation cannot only drive but also inhibit the growth of regulatory capitalism, at least for a certain period. Third, competition between different self-regulatory strategies may serve as an impediment to the diffusion of regulatory capitalism. In this case, both risk management and BBS are businesses' self-regulatory strategies and the growth of either approach increases the diffusion of regulatory capitalism. Nevertheless, since risk management is a key embodiment of the new order of regulatory capitalism, the existence of the relatively traditional BBS to some extent makes the diffusion of regulatory capitalism slower in the U.S. coal mining industry.

Another issue is worthy of mentioning here. As discussed in the case of Blacksburg Coal, the trigger of its adoption of BBS was two fatal accidents in 2004. This again illustrates the point raised in the last chapter: driving up standards in the industry is also a reactive process to disasters. Therefore, based on all of the findings in Chapters 3 and 4, the analytical framework needs to be modified (Figure 4.4). This modified framework, synthesizing the proactive and reactive sequences and adding the element of a competing model, can provide stronger explanatory power than the original Braithwaite and Drahos framework. It has the power to explain not only the driving forces, but also the counter forces of a modeling process.

Chapters 3 and 4 have shown the modeling process of risk management in the U.S. coal mining industry and the barriers that prevent this model from

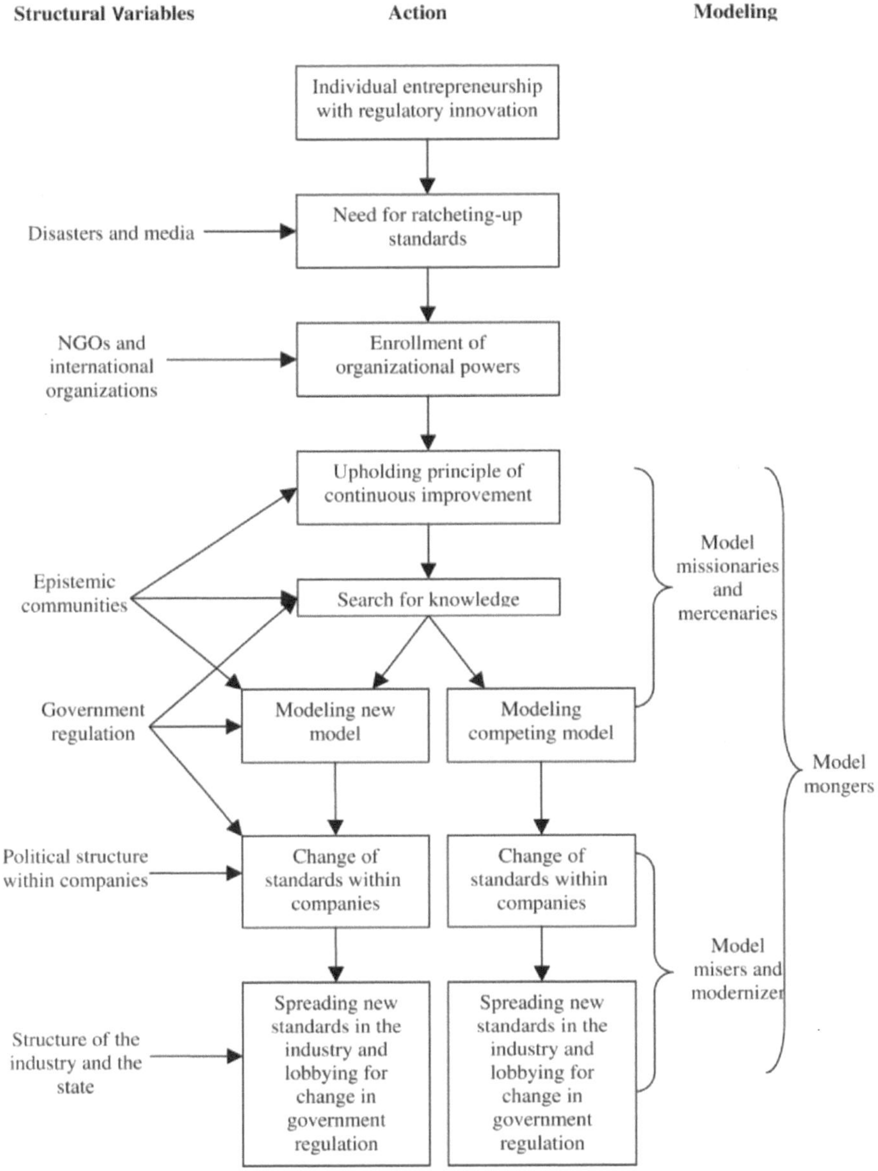

Figure 4.4 Modified analytical framework: driving up standards in the industry.

fully diffusing. The focus of the discussion in these two chapters has been at the level of companies and the industry. As Chapter 2 discussed, risk-based governance includes two interrelated components: enterprises' internal risk management and government risk-based system regulation that mandates or encourages the establishment of internal risk management systems. Although government risk-based system regulation is not a sufficient factor that leads to the growth of enterprises' internal risk management, government mandates through statutes and regulations can bring the model of risk management to the attention of more in the industry. As this chapter discussed, one of the barriers to modeling risk management in the U.S. coal mining industry is the prescriptive regulatory climate. By contrast, in Australia, the extensive use of the risk management model across the coal mining industry is, to a large extent, the result of the shift away from prescriptive regulation toward risk-based system regulation. In the U.S., although the basic regulatory structure remains highly prescriptive, efforts have been taken by the national government regulator—MSHA—to move toward risk-based governance. However, MSHA's efforts were not successful. Why did MSHA's efforts not obtain similar outcomes as its Australian counterpart? This is the question to be answered in the next two chapters.

5 Modeling Risk-Based System Regulation in Australia

INTRODUCTION

To better answer the third sub-question (what are the barriers keeping the U.S. government from moving toward risk-based system regulation?), it is helpful to examine first how and why the Australian coal mining safety regulation has successfully shifted away from prescriptive regulation toward a risk-based system regulation.

Based on the analytical framework, the research finds three factors that help explain the success of regulatory reform in Australia. Following a modeling explanation, the success resulted from (1) strong imitative pressure from general OHS regulation; (2) the existence of strong model mongers, missionaries, and mercenaries. More specifically, model mongers include senior government inspectors, mine operators and managers, as well as union check inspectors. They are the major drivers of the move toward the risk-based system regulatory model because they have the cognitive capacities to understand the value of the model, and they also see perceived benefits. In order to achieve an outcome that can be accepted by all parties, these model mongers engage in dialogue with a contending group of model mongers who support more prescriptive regulations. In this process, model missionaries and mercenaries play a significant role in helping different parties to achieve shared understandings. The dialogues among model mongers, missionaries, and mercenaries are supported by the existence of institutionalized webs of dialogue, the third variable explaining the success of regulatory reform in Australia. The formation of webs of dialogue, which are often exercised through tripartite mechanisms, is influenced by industrial relations, cooperative culture, and a parliamentary political system.

In Australia, occupational health and safety regulation is handled at the state level. There are three "mining states": Queensland (Qld), New South Wales (NSW), and Western Australia (WA). Most coal fields in Australia are located in Queensland and NSW, while WA is dominated by metalliferous mining. Although metalliferous mining will also be mentioned, this study focuses on coal mining safety regulation. Most of this chapter's discussion is related to regulatory reforms in Queensland and NSW. Before examining the influence of the three independent variables, a description of the regulatory reform in Australian coal mining safety regulation is provided.

SHIFT TOWARD A RISK-BASED SYSTEM REGULATION

The forces that drove the change in mining safety regulation in Australia go back to the Robens Report published in 1972 in the U.K., which gave a comprehensive review of the occupational health and safety regulation at the time. The Report concluded that highly prescriptive regulation has limits in improving occupational health and safety. For example, it creates massive detailed laws, and problems "falling between the cracks"; it encourages a minimum compliance mentality, diminishes employers' responsibility for safety, and imposes unnecessary costs on companies and regulators (Gunningham, 1999; 2004). To deal with the problems of prescriptive regulation, the Robens Report recommended that (1) primary responsibility for safety should lie with those who create the risks, and, therefore, general duties of care should be introduced for all people at the workplace including both employers and employees; (2) alternative regulatory standards to specification standards should be adopted, such as performance-based and systems-based standards (*ibid.*).

Based on the recommendations of the Robens Report, the U.K. Health and Safety at Work Act was enacted in 1974. The Act was a single statute regulating workplace safety in all industries, including coal mining. It imposed a general duty on employers to provide safe and healthy workplaces. This Act also took a risk-based approach to health and safety. On the one hand, the requirement of a general duty of care implies that employers need to develop risk-based management systems to fulfill their responsibilities. On the other hand, the Act requires the regulator to take account of the potential risks and the health and safety management performance of duty holders (*ibid*). The Robens Report and the U.K. Health and Safety at Work Act created a "Robens model" of regulation.

Because of the historical linkage between the U.K. and Australia, this Robens model of regulation had "immediate and direct" impacts on Australia.[1] During the 1980s and 1990s, general OHS regulations in NSW and Queensland gradually embraced the general duty of care and more performance- and systems-based standards. However, this change in general OHS regulation had limited influence on mining regulations until the late 1990s because mining was under legislation separate from general OHS legislation, which covered all other industries.

This situation has changed since 1999. In that year, Queensland introduced the Coal Mining Safety and Health Act 1999 (Qld) (CMSHA 1999 [Qld]) and the Mining and Quarrying Safety and Health Act 1999 (Qld) (MQSHA 1999 [Qld]), which applied to coal operations and metalliferous mines and quarries, respectively. Although these two mining statutes are still separated from general OHS legislation, they are similar in structure and content to general OHS legislation. Different from Queensland, regulatory reform in NSW created a hybrid regime, in which the legislature enacted general OHS legislation—the Occupational Health and Safety Act 2000 (NSW) that applied to all workplaces including mining. At the same time, mines are subject to two additional mine-specific statutes—the Coal Mine Health and Safety Act 2002 (NSW) (CMHSA 2002 [NSW]) and the Mine Health and Safety Act 2004 (NSW) (MHSA 2004 [NSW]), which covers metalliferous mines and quarries (Gunningham, 2007, 18–19).

These new mining legislation introduced two new types of standards: general duties and risk management systems-based standards. According to Gunningham (2007, 21), the advantage of general duties is their "all-encompassing character." Because of this comprehensive character, employers have to continue improving safety as long as it is "reasonably practicable."[2] Although the two mine-specific statutes in Queensland do not include general duties provisions, they require employers to manage risk to an "acceptable level." This risk management requirement can achieve results similar to general duties standards (*ibid.*).[3]

Risk management systems-based standards in the new mining legislation are not only general risk management principles, but also requirements to establish more detailed hazard management plans for specific hazards, considering that mining operations are high hazard workplaces. For instance, section 32 of CMHSA 2002 (NSW) requires employers to establish major hazard management plans according to regulations. The regulations specify that major hazards include slope stability, surface transport, underground transport, strata failure, in-rush, fire and explosion, dust explosion, explosives, and airborne dust (Gunningham, 2007, 28). In Queensland, a principal hazard management plan is required under section 62 and 63 of CMSHA 1999 (Qld). This plan must include standard operating procedures and other measures to control risks (*ibid.*). Moreover, the major or principal hazard management plan must define the triggers of action, what actions are to be taken, and who is responsible for the actions (Hopkins, 1999).

The general duties and risk management system standards are closely related. As previously stated, the virtue of general duties is their comprehensive character. However, these general obligations can be too vague. When regulation moves away from specification standards toward general duties, the question arises of how to tell whether an employer has adequately committed to its duties of care. A way for employers to demonstrate their duty of care is to put risk management procedures in place. This role of risk management as a way to demonstrate duty of care has been highlighted by the court's interpretation in Australia.[4]

From this description, it can be seen that the shift of the regulatory framework toward risk-based system regulation is clearly a learning process through modeling.[5] The rest of this chapter will explain why the risk-based system regulation has been successfully adopted in Australian coal mining safety regulation.

MODELING RISK-BASED SYSTEM REGULATION

Following the analytical framework, this section examines three variables explaining the success of the Australian government's effort to move toward risk-based system regulation. The three explanatory variables are: (1) strong imitative pressure from general OHS regulation; (2) the existence of strong model mongers, missionaries, and mercenaries; and (3) the existence of webs of dialogue.

Imitative Pressure

As previously discussed, modeling is first an imitative process. According to Max Weber, some people open their umbrellas because of rain, whereas others may open theirs by following what others do (Levi-Faur, 2005). Powell and DiMaggio (1991) also recognize imitation as one of the isomorphic pressures for institutional change. The description of the shift away from traditional prescriptive legislation to a new-generation regulatory framework in the last section showed that the statutory changes in the Australian mining sector clearly were imitations of the development of general OHS legislation. As Braithwaite and Drahos (2000, 538) note, modeling is patterned in various aspects. Related to wealth and power, the ones who are powerful and wealthy are usually modeled by those who are not. However, modeling may not necessarily be associated with power or wealth. It often occurs between actors or fields sharing similar attributes. The most similar field in relation to mining safety regulation is certainly general OHS regulation.

When those in the mining sector in Australia considered the change in the regulatory model in the 1990s, general OHS regulation was the first area they turned to look at, and the model was already there. For instance, the Occupational Health and Safety Act of 1983 (NSW) (OHS Act 1983 [NSW]) introduced general duties and system requirements in general OHS regulation in NSW (Gunningham, 2007, 15). The imitative pressure from OHS legislation was mentioned by many informants. As one noted:

> The general argument is OHS legislation has moved more to sort of a risk management approach. So there is an argument that mining also needs to move to a more risk management approach. . . . They try to make mining legislation look more like health and safety legislation.

The imitative pressure came not only from Australian OHS regulation, but also from OHS regulations in other countries. As Chapter 2 showed, the ideas of OHS management systems and risk management have been gradually adopted in general OHS regulations in many European countries since the 1970s. The Robens-style regulation in the U.K., the EU risk management framework (the Framework Directive [89/391/EEC]) passed in 1989, and internal control systems mandated in Scandinavian countries in the 1990s are examples (Frick et al., 2000). One informant said:

> It is an international trend that has happened generally through health and safety legislation in other developed countries but not in the U.S. . . . The changes in OHS legislation and mining legislation that we are talking about are not unique to Australia. Those sorts of things are part of a more general shift. . . . When we remodel our legislation, we do look at what other countries are doing because if someone has invented the wheel in other places, why would I go on to reinvent it?

When the state governments in Australia considered a change of mining legislation, they saw that the whole regulatory framework in the general OHS area had moved to the risk-based system approach. In other words, "the direction was already apparent."[6]

As Braithwaite and Drahos (2000) argue, modeling is not only an imitation, but also involves cognitive interpretation and transforming information into symbolic form. Without symbolic interpretation, a model cannot become a guide for future action and diffuse to a larger scale beyond a one-to-one environment. This cognitive process is particularly important for those people who actively push the model to a higher level. A follower may simply do what others do. A leader, however, has to understand the model's implications because they are actually the persons who conduct symbolic interpretations. In addition, unlike person-to-person or firm-to-firm modeling, a modeling process at the government policy level involves more parties that have diverse interests. Therefore, the actual legislation and regulations—outcomes of symbolic interpretation—need coordinated efforts by different parties that may have different interpretations of a new model. The detailed discussion of this modeling process appears below, focusing on several key actors (e.g., model mongers, missionaries, and mercenaries) and their cognitive interpretations.

Model Mongers, Missionaries, and Mercenaries and Their Cognitive Interpretations

Before discussing the role of these key actors in the modeling process, it is necessary to give a brief description of the micro-macro dynamics during the changes in mining statutes and regulations. The purpose is to show readers circumstances in which key actors exercise their influences.

Micro-Macro Dynamics of Regulatory Change

According to Braithwaite and Drahos (2000), in the reactive sequence of regulatory change, individual entrepreneurship from businesses or NGOs is less important. More important is that key individual entrepreneurs who have "direct organizational control of [the] state's capacity" act in response to disasters that provoke the mass public to demand change (561). These individual entrepreneurs are strategic model mongers. They have a number of regulatory models in hand and wait for a disaster to open a policy window. When the window opens, they pull out a persuasive model from their desk as a remedy to the crisis.

This process partially applies in the case of the change of mining legislation in Australia. It is correct that the model mongers who had "direct organizational control of [the] state" (senior government inspectors in this case) are more important than the individual entrepreneurs in the mining industry or NGOs. However, disaster was not the only trigger of government action. Changes of government can also open a policy window. In addition, model mongers in this case did not have multiple models as Braithwaite and Drahos (2000) portrayed. This is because of the isomorphic influence of general OHS regulation as the last section described. The direction was already clear when the government wanted to change the mining legislation. All parties agreed on the general direction. The disagreement was over "how much prescriptive regulation should be put in the law and how much should be taken out."[7] This reflects the power of the model that Braithwaite and Drahos (2000) suggest. Once it is put on the table, the model sets the framework for debate.

According to information I collected from interviewees and documents, in Queensland, this risk-based system regulatory model had been discussed for many years before 1990; being influenced by the Robens report in the U.K. and the change in general OHS regulations. However, the Queensland government did not do anything until the Labor Party took over the government in the early 1990s. According to informants, the previous conservative government, which had been in power for 25 years, had lost legitimacy. There was little progress in many areas. When the Labor government came into office, there was "massive reform of everything." In the words of an interviewed inspector, the new government came in "to sweep out the sins of 25 years of conservative party domination . . . bribery and corruption."

In 1991, this new Labor government decided that "all legislation should be reviewed."[8] The Department of Mines and Energy was responsible for reviewing mining health and safety legislation. In 1992, a tripartite Legislation Review Group was established, including members of the mine inspectorate, unions, and the industry (Billingham, 2000). The government initially wanted to have one piece of legislation that applied to both coal and metalliferous mines. However, "the parties of coal mining and non-coal mining did not speak to each other."[9] This is partly because of their different cultures and traditions, and partly because the two sectors have separated unions. One is for coal mining and the other is for metalliferous, quarry, and stone mining. They also "fight each other for membership."[10]

As a result, the government decided to have two different working groups to look separately at the legislation for coal and for metalliferous mines. Each working group included representatives of the regulator, the unions, and the mine operators. According to an interviewee, all people involved in these working groups were "practical people and mining people; lawyers were not allowed." When he explained why lawyers were excluded, he said "we are trying to improve safety in the working place. Mining people know what is going on in the working place, not lawyers."

Two legislation review groups were formed first to develop the proposed legislation. Subsequently, almost the same people constituted regulation review groups. The development of new regulations started in 1994. In terms of the review of coal mining safety regulation, two review committees were established to discuss surface and underground mining regulation. Unfortunately, according to Bancroft (2000), members of the underground committee could not agree on some issues, and some members left. The review was suspended between August 1994 and February 1997, waiting for the findings of the inquiry about the Moura No. 2 mine explosion, which occurred in August 1994.

When the disaster inquiry results came out in 1996, the two working groups joined together again, with clearer direction based on the recommendations of the Moura disaster inquiry. A new committee was formed for developing underground coal regulation. As an informant who was involved described, they did not change principles, such as non-prescriptive legislation and the need for people at mine sites to manage risks. However, particularly for coal mining, they introduced more prescriptive requirements that mines must have specific management plans for principal hazards that may cause major disasters.[11]

Thus, the changes in mining statutes and regulations were undertaken mainly through small legislation and regulation review groups that consisted of representatives from the government, the industry, and unions. The representatives in the review groups are actually model mongers of the new regulatory model. The next sub-sections will explain why they could become model mongers and how they interpreted the risk-based system regulatory model.

Model Mongers

According to Braithwaite and Drahos (2000, 579), the model mongering process is the most important one for weak groups to acquire regulatory change because it can strategically produce recognition of an identity crisis that "opens cracks in hegemonic structures." There are three major groups of model mongers in this study: senior government inspectors; mine operators and managers; and union check inspectors.

Senior Government Inspectors

Senior government inspectors are the most important model mongers in this case. As the 1997 NSW *Mine Safety Review* stated, the "inspectorate, and in particular the Chief Inspector of Coal Mines, was seen as having been a driving force in encouraging companies to take on a risk management approach" (Acil, 1997, 33). Many informants also mentioned the crucial role of senior government inspectors.

According to Bandura's (1986) social cognitive theory, observers have to be exposed to a new model first before beginning the process of learning by modeling. However, even though exposure to a new model is necessary, it is not sufficient for them to adopt the new model in practice. Observers also have to have the cognitive capacity to transform information about the model into symbolic form that serves as a guide for future action. Moreover, some additional factors, such as environmental pressures and incentives, also influence observers' decisions on whether they will act on what they have learned.

As mentioned, the most important symbolic document that brought the risk-based system regulatory model to Australia was the Robens Report in the U.K. All interviewed inspectors noted that they were aware of the Robens Report right after it came out. For example, a senior inspector said:

> In the department, we realized that there was a major reform in the U.K., the Robens Report, which talked about bringing health and safety into one piece of legislation. It also talked about risk management and, most importantly, pointed out the fact that the duty relies on the people who create them. It is not government that creates the risks. It is people who own the operations and workers themselves.[12]

According to Braithwaite and Drahos (2000, 581), modeling is defined as actions that constitute a process of displaying, symbolically interpreting, and copying conceptions of action. Therefore, modeling includes the process of both model supply and adoption. The influence of the Robens Report can be viewed as displaying conceptions of action.

Exposure to a new model, however, does not necessarily lead to changes in behavior if the observer does not have the cognitive capacity to perceive the value of the model. Government inspectors in Australia have the cognitive capacity to understand the risk management model because they all have experience being mine managers. All mine inspectors in Australia are required to have mine manager certificates. Many senior inspectors had their early careers as mine managers. Therefore, they are experienced in comprehensive management of all aspects of mining operations. They also understand how important risk management systems are to mine safety. Therefore, it is easier for them to understand risk management and systems-based regulatory approaches. As Rogers (2003) shows, perception is very important for diffusion. People who perceive less uncertainty will be more likely to adopt a model than those who perceive a model as having a lot of uncertainty.

Although he talked about mine managers in his American mining companies, an informant's explanation indicates why mine managers in general find it easy to accept the risk management model.

> Risk management is straightforward to senior management because they do risk assessment all the time. What do they do every time they open a coal mine, decide how to spend their money on a piece of equipment, or whatever. To some extent they are probably more attuned to it than anybody.[13]

Australian government inspectors' experience in risk management was also demonstrated by an informant who had experience training them. He talked about his impression of government inspectors:

> The regulators in Australia have always been mine managers, ex-mine managers, who become regulators. . . . Mine manager persons were usually mine engineers or other persons who came out through the ranks [and got] a license to manage mines and therefore they have been there at least five years. . . . Usually people are just tired of the pressure in mines and find a job as regulators. . . . Regulators are selected very carefully, the regulators are normally very experienced mining managers. They have five mines to take care of. So they regulate and work with the five mines . . . everything, all the areas. It was not hard for them to make the change to risk stuff. I taught the regulators about risk stuff. They went to a course about risk assessment. They come with knowledgeable risk information.

Besides exposure to the risk-based regulatory model and their cognitive capacity, some additional factors have stimulated government inspectors' actions to change regulatory policy in Australia. One factor is environmental pressure, such as the isomorphic pressure from general OHS regulation. In addition, they have incentives to adopt the model, such as anticipated benefits.

Two anticipated benefits are particularly important for government inspectors. First, requiring employers to establish internal risk management put primary responsibility for workplace safety back on employers. Before the regulatory reform, many coal mining companies activities were subject to regulators' approvals. As such, "risk ownership was on the government, not on the person who created the risk."[14] In fact, when some disasters occurred in the 1990s such as the Moura No. 2 mine explosion in Queensland and the Gretley inundation in NSW, inspectors had been criticized for being liable for the disasters because they did not properly enforce regulations and made questionable approvals. An interviewee, who was a general manager of mine inspectors in Queensland, portrayed the situation that inspectors faced after the Moura No. 2 mine disaster occurred:

> I was in the situation of having inspectors who were under extreme stress both because they inadequately enforced regulations, and in court under prosecutor examination, being cross examined about whether the failure to enforce regulation was the cause of the disaster. My view was that this was a very unreasonable suggestion to be making, that somehow the inspector should have been preventing this. It is implicitly saying that you make mines safe by enforcing regulation. If a mine is forced to comply with regulation, there cannot be a disaster. If there is a disaster, it comes from the failure of enforcement. It takes responsibility of safety operation away from mine operators, and puts it on the third party, the inspector or regulator. This is fundamentally wrong. We need a system that says compliance with guidelines, rules, and regulations is a minimum but not sufficient condition to ensure safety. The responsibility for safety has to be on mine managers.

This shift of responsibility back to employers is one of the reasons why inspectors are in favor of risk-based system regulation. This government's intention can be demonstrated by the Construction Forestry Mining Energy Union (CFMEU)'s submission to the NSW *Wran Mine Safety Review*, 2005. CFMEU critically noted that the Department of Primary Industries "has its own agenda of seeking to divest itself of many of its regulatory responsibilities" (Wran and McClelland, 2005b, 18).

Second, the adoption of a risk-based system regulation can help government inspectors to do their jobs in a more cost-effective manner. In an environment of neo-liberal governance, government encounters increasing constraints of public finance. Therefore, governments have to seek more cost-effective ways to regulate. Compared with regulators in other countries, particularly MSHA in the U.S., Australian regulators are relatively small in terms of both budgets and human resources. Therefore, they do not have the capacity to check every single aspect of mines as MSHA inspectors do in the U.S. When regulations contain thousands of pages of specific standards, inspectors have to go through the large number of standards one by one when they enter mine sites. According to an evaluation of MSHA's inspection program in the U.S. conducted by ICF Consulting, the average time per inspection in 2002 was 108.1 hours (about two and a half weeks) for an underground coal mine (ICF, 2003). For a large mine site, it may take two or three months for an inspection. As an Australian informant explained:

> It is recognition that we don't have the capacity to do it. We don't have the resources and money and people. We could not afford MSHA in Australia. That would be too expensive. . . . In our government departments, there were chief inspectors who drafted that idea [risk-based

> system regulation] and they run very actively because they recognize the value of the process but also sort of make their job easier.

By shifting more responsibilities to employers, regulators would certainly do much less work than before. Government does not have to approve companies' activities any more. It is the mining companies' responsibility to demonstrate that they do the right thing.[15]

Therefore, Australian regulators thought "if management could put safety management systems in place that actually worked," they could get "better outcomes for less resources." [16] According to one scholar, this is wishful thinking and may not work exactly the way they hoped in practice.[17] This notion reflects that the anticipated benefits influence observers' action, but not necessarily the actual benefits.

As a result of their cognitive capacity and the anticipated benefits, senior government inspectors put the risk-based system regulatory model on the table when the government decided to change existing regulatory structures. An interviewee gave a summary of this complex process:

> If a disaster occurs, there is pressure politically on the minister to do something differently. Those very experienced senior mining people, those people brought in what the Robens Report said. If we change, we have to switch to this way. *The process I think why it happened fairly quickly in Australia is that senior regulators are very experienced mining people who recognize that the change of regulation is required to reduce disasters.* They looked at the Robens Report and Pipa Alpha disaster. *Those recommendations made sense personally, and they justify the change to politicians* [emphasis added].

Mine Operators and Managers

In addition to government inspectors, another group of model mongers of the risk-based system regulation is mine operators and managers. When the regulatory reform had been put on the political agenda by the government, these people from mining companies tried to "shape the legislation in a way that gives them flexibility to deal with safety in the way they want to deal with it."[18] In other words, this new regulatory model had an anticipated benefit for them as well—the flexibility to deal with different situations in different mines. For example, if a prescriptive standard requires that the distance between roof bolts has to be 8 feet, mine operators and managers may say "why 8 feet? It may be 10 feet. There are different situations in different mines, so you have to look at risks and determine how to manage the risks."[19] In general, the management of mining companies in Australia wants "resilience" and "individualized management structure."[20]

However, not all mining operators and managers are model mongers of the risk-based system regulatory model. Some mining operators also support a prescriptive approach because they perceive more benefits than from the risk-based system regulatory model. First, they see the prescriptive approach as easy to comply with. The only thing they have to do is to comply with the law, and they "need not have to think about things."[21] Second, they see the general duties standards as being too broad. If an employer has breached the law, any individual in a management position can be liable unless a defense can be established (Ellison, 2005). Therefore, "they don't like a Pandora's box being opened."[22] Third, they are particularly sensitive to the liability imposed on employers to demonstrate their duty of care in criminal prosecution. Just because of the last two concerns, three companies—Xstrata, Centennial Coal, and Coal Operations Australia, Ltd (COAL)—challenged the convictions of the New South Wales Industrial Relations Commission in court.[23] This litigation has the potential to overturn the basis of OHS legislation in NSW. This case is currently under review.

Union Check Inspectors

In addition to government inspectors and the majority of mine operators and managers, another group of model mongers of the risk-based system regulation consists of some union officials. This is surprising because unions in general are typically model mongers of traditional prescriptive regulations for at least three reasons. First, unions perceive that prescriptive regulation has more anticipated benefits than the risk-based system regulatory model. This is because black and white standards make it easier for them to identify breaches of law and, therefore, they can hold employers accountable. By contrast, risk management is vaguer and more difficult to understand. It is not easy to tell whether an employer has breached the law or not.

Second, unions also see anticipated consequences of risk management. According to Bandura's (1986) social cognitive theory, people's behavior is motivated by anticipated consequences of their actions. The anticipated consequences, in turn, stem from observed consequences of previous actions. Unions have seen the evidence that management has failed to effectively manage risks. This has been demonstrated in many government documents, such as the *Wran Mine Safety Review,* 2005, in NSW. The *Review* found a "disconnection" between "corporate management systems/plans and their translation to a working level at the 'coal face'"(Wran and McClelland, 2005b, 49). Moreover, in the Moura disaster, the mine did have risk management systems in place. However, the systems were not managed properly. This is why unions advocated the more prescriptive principal hazard management plan in Queensland, as mentioned earlier.

Third, unions tend to support the prescriptive approach because of their members' cognitive capacity. Unlike mine managers, workers find it more difficult to understand complicated risk management systems. Therefore, they prefer straightforward prescriptive standards. If a regulation says to keep 8 feet between roof bolts, they just measure 8 feet.

While unions generally favor a more prescriptive approach, this does not mean that they are completely opposed to the risk-based system regulatory model. First, they like the duty of care approach because it places more obligations upon employers. For instance, in the case of the three companies' appeal in NSW, CFMEU has already threatened to go on strike to protect duty of care provisions in OHS legislation (Ellison, 2005). Second, some union members see risk management as an opportunity to participate in developing mine safety management systems.[24]

Most important, there are also model mongers of the risk-based system regulatory model in unions: the union check inspectors. A union check inspector I interviewed in Queensland indicated that the reason why unions also supported the change to the risk management approach in general was because of the check inspectors' influence. In Australia, besides government inspectors, there are also union check inspectors elected by the workforce. Like government inspectors, these union check inspectors have the authority to access all coal mines, stop production if any emergent hazard is identified, ask operators to review risk assessment and revise management processes, and investigate accidents. According to an interviewee, they are "intelligent people with incredible practical knowledge." Therefore, these union check inspectors also have the cognitive capacity to perceive the value of risk management. As a respondent noted:

> They [check inspectors] went out to learn when risk management just began. They know as much about it as anyone else in the advisory council. How it should work, how it does work, and how it doesn't work. So they embrace it.

These union check inspectors have the cognitive capacity mainly for three reasons. First, they must be qualified individuals to be elected as union check inspectors. There are only three union check inspectors in Queensland. Second, their experience as inspectors provides them opportunities to constantly evaluate the root cause of accidents. Third, when they are elected as check inspectors, unions provide them many opportunities to learn the best practices. An interviewed union check inspector said:

> When you are qualified to get to that position, unions put in a lot of training, bring you to NIOSH, MSHA, Germany, and England. Unions spent a lot of money to make sure the safety inspectors elected have the leading knowledge, so you are respected when you go to different places. If you don't have the knowledge, you won't be respected in the industry.

Moreover, these check inspectors were often designated as union representatives in the working groups to develop statutes and regulations.[25] According to the union check inspector I interviewed, two out of the three check inspectors in Queensland served as union representatives on the legislation and regulation review groups. This provides them the opportunity to exert influence.

In short, the model mongers of the risk-based systems regulatory model are primarily government inspectors and mine employers and managers, but they also include some check inspectors from unions. By contrast, there is a contending group of model mongers who support more prescriptive regulations. They are mainly from unions, but also include some employers or managers. The existence of this contending group with vested interests in preserving existing practices is the largest obstacle to adopting a new model. Under this circumstance, valued outcomes cannot be achieved unless doing so provides some benefits for all parties. The symbolic interpretation of the new model (e.g., legislation and regulations) must involve some "interdependent contingencies, which tie people's benefit to progress toward common goals" (Bandura, 1986, 159). This is exactly what has happened in Australia. The next sub-section will show how model mongers from contending groups interpreted the risk-based system regulatory model.

Symbolic Interpretation among Model Mongers—A Politics of Give and Take and Justifying by Expert Knowledge

According to the theory of modeling, statutes and regulations are actually the outcomes of symbolic interpretations that provide guidance for future actions. People's interpretation is often based on their interests. In this case, unions tend to interpret a new regulatory model into a more prescriptive style, whereas most employers and managers prefer more flexibility.[26] As a result, new mining statutes and regulations are a compromise between two contending groups of model mongers.

On the one hand, the law provides operators a certain level of flexibility to deal with particular circumstances in their operations by embracing more generally phrased standards instead of specific standards. At the same time, more responsibilities are imposed on operators: a duty of care is the price of flexibility. On the other hand, unions are worried that employers may try to cut corners when they hear owners say that they are going to adopt a different approach because their site is special.[27] Union officials argue that if the employers are allowed to have flexibility to put risk management in place, they have to be held accountable for any alternatives they choose.

Consequently, the new mining legislation creates a three-tiered framework: legislation, regulation, and code of practice (recognized standards). The legislation only imposes general obligations on employers to control risk with little prescription. Regulations, which are subsidiary legislation, set more detailed prescriptive requirements on the particular hazards that need to be controlled and what procedures are necessary. For example, Coal Mine Health and Safety Regulation 2006 (NSW) specifies that major hazard management plans need to be developed for the following hazards in underground mining: slope stability, surface transport, underground transport, strata failure, in-rush, fire and explosion, dust explosion, explosives, and airborne dust (Gunningham, 2007, 28). When determining how to implement the legal requirements at mine sites, employers have to have agreements with labor. If they cannot reach agreement, the legislation requires that employers have to adopt guidelines, codes of practice, or recognized standards developed by the inspectorate joined by the industry and union representatives. Otherwise, employers have to show specific justifications if they choose alternatives.

Therefore, model mongering is not only a politics of empowerment as Braithwaite and Drahos (2000) suggest; it is also a politics of give and take. Since different parties have differing interpretations of the model, it is the politics of give and take that determines the outcome of new mining legislation—a symbolic form of a new regulatory model—in which all parties have some benefits.

The politics of give and take often occurs when discussing issues related to the larger regulatory framework. According to some informants, when a more detailed regulatory issue is involved, particularly in developing a code of practice or recognized standards, it is less about the matter to get an agreement or compromise, and more about the matter to justify it. In other words, when people cannot agree on a particular issue, the rules developed need to be the "best outcome" based on scientific information. An important way to justify rules is use external experts to provide the "best" knowledge about the issue. For example, operators and unions for many years disagreed on what a drug test means and how it should be conducted. Operators say if you use any drugs you should not work on the mine. There is risk. The unions say that if somebody smoked marijuana a week ago, when the employee comes back to work, it does not matter. To resolve this problem, the advisory council (a tripartite advisory body that will be discussed later) invited the best experts in the country to talk about the issues, such as the effect of drugs on the ability to drive, how to test for drugs, and how the police do drug testing.

As this example shows, when different parties' opinions are polarized, experts bring different parties' positions closer if not completely together. This is the role that model missionaries and mercenaries played in the modeling process in Australia.

Model Missionaries and Mercenaries

Model missionaries and mercenaries are those scholars and consultants who believe in risk management, and who also make profits by providing consulting services for government, companies, and unions. Model missionaries and mercenaries are important for the modeling process because modeling involves acquiring knowledge, processing information, transforming it into symbolic form, and implementing the model in practice. During this process, model missionaries and mercenaries can distribute knowledge and help different parties to achieve shared understandings because they are perceived as independent from self-interested stakeholders. In short, model missionaries and mercenaries can shape people's cognition. They can exercise their influence through a number of channels, such as making presentations in regular health and safety conferences, providing training to inspectors or companies, and participating in accident investigations and mine safety reviews for the government. Some examples of these influences are explained in the following paragraphs.

Jim Joy, a well-known scholar and consultant on risk management, provides training for government inspectors, mining companies, and union members. Starting in the 1990s, the University of Queensland introduced courses on risk management led by Joy. An inspector talked about Joy's influence:

> Before Jim Joy some people came in to train us about risk management. But they were not very influential. When Jim Joy came along, he established a postgraduate risk management course at the university. Inspectors and union people all start taking this course . . . the way he has driven risk management to the university has been phenomenal . . . *the best course I have ever been to is Jim Joy's*. . . . He has managed to get to everybody. He tries to convince us this is the way to go. . . . When taking Jim Joy's courses, many people said "this is the level of risk management competence that people in the industry should have"[28] [emphasis added].

Most important in this example is that the participants in this training course are not only government inspectors but also people from the industry and unions. An interviewed union check inspector indicated that he started to join Joy's training in 1991. It is not hard to imagine the influence when people from all major parties come out of the same classroom.

A second example of model missionaries and mercenaries' influence is their frequent participation in accident investigations and mine safety reviews for the government. According to Bandura (1986, 116), the environment's influence on people's behavior often goes through the "intervening influence" of human thought. These investigations and reviews are actually the process in which people perceive what happened in the past and develop guidance for future actions. They have a great impact on people's cognition because reports are distributed to all stakeholders. In Australia, the selection of group members for government mine safety reviews is based on a tripartite principle. Government, operators, and unions each nominate one expert. These experts, although they may come from different areas, usually can agree on major principles. This is very helpful in creating shared understandings among different parties. An interviewed scholar recalled his experience with the *Wran Mine Safety Review*:

> When we did that review, there was a clear view among all of us that systems were generally a good idea. Jim Joy was the management nominee. I was brought in by CFMEU [Construction, Forestry, Mining and Energy Union]. There was a guy from the offshore oil authority who was the government nominee. We agreed on about 85% of everything. One was that system [managing safety] was good. That is the thing we had to look at. We did identify some problems with systems, such as inadequate feedback loops. We took some of the union criticisms of systems on board, and some evidence was given.

His description was reinforced by the text in the *Wran Mine Safety Review* Report. It shows that three experts were appointed as advisers for the reference panel for the *Wran Mine Safety Review*, headed by the former premier, Neville Wran, AC, QC (Wran and McClelland, 2005b, 9). The three experts were Jim Joy (Professor of Mining Safety and Director of the Minerals Industry Safety and Health Centre at the University of Queensland), Professor Michael Quinlan (School of Organization and Management, University of NSW), and Peter Wilkinson (Project Manager, National Offshore Petroleum Safety Authority Implementation Team) (*ibid.*).

The *Wran Mine Safety Review* Report is also consistent with the informant's observation that there was an agreement among experts on the general direction of regulatory reform:

> The Review considers the consensus of expert opinion favors a shift to risk-based legislation, but with the retention of prescriptive regulation in particular areas (e.g., where the safety factor of the risk is uncertain and a careful threshold is required, such as mine gas levels). (Wran and McClelland, 2005b, 22)

It can be seen that these experts are not necessarily specialized in mining. Quinlan is a scholar in general management, and Wilkinson is a specialist in offshore safety. This kind of mechanism can help to draw lessons from other industries to mining. As Wran and McClelland (2005b, 9) note, the role of these expert advisers is to provide comment and analysis, "drawing on their respective areas of expertise and experience" including "comparative information and insights on matters from other industries and jurisdictions."

Different parties are more likely to reach shared understandings with the involvement of such model missionaries and mercenaries. This is because, unlike those coming directly from the industry or unions who tend to be predisposed toward their own interests, these experts take more independent roles in the review. Although they are nominated by different parties, they are not the agents of those parties. This independent role makes people tend to believe what experts say. An informant said:

> They know me and they know my work. We have to pick an expert. We probably pick him because we sort of know what he does. We trust him more than we might trust someone else. I would say that I do things like this. I will do my professional job. I promise you nothing else. Because you nominate me, it doesn't mean I just do whatever things you want. I would get that principle because I am not valuable to anybody if I just start writing things for other people.[29]

Saksvik and Quinlan (2003, 38) also indicate that the OHS consultant and bodies are one "important contributor to promoting a 'systems' approach in Australia" and the "capacity of consultants was enhanced by their status as independent brokers."

This type of collaboration among model missionaries and mercenaries through review groups on accident or government policy itself is a way of diffusing knowledge and ideas. An interviewed consultant described how he became aware of Neil Gunningham's systems-based regulation. When he was involved in the review of the Queensland mine inspectorate in 2004, he said that he worked with Andrew Shaw, an independent specialist on occupational safety and health. Shaw is very familiar with Gunningham's work and introduced it to the consultant.

In general, the risk-based system regulation can be modeled through in Australia because the model mongers, missionaries, and mercenaries discussed are the "right people."[30] Most important, the model mongers—the representatives nominated by stakeholders working in small legislation and regulation review groups—are people who have foresight.[31] An informant said:

> We selected groups of stakeholders and asked them to nominate people for our review groups. Then they nominated the right people. They were very good people on all sides. They were engaged and knew the job . . . *we had the right people. That was the only way we could go ahead.* Certainly on the coal mining side, there were a lot of powerful battles. They got through them [emphasis added].

Even unions, which in general are more favorable to a prescriptive approach, nominated check inspectors to the review groups. As previously stated, these check inspectors are experienced in risk management. They also embrace it because they know the necessity of risk management based on their experience.

Despite the involvement of those right people, modeling still may not succeed without the support of the structural environment. For example, it is difficult for the representatives of a legislative working group to make good judgments independently if they are constrained by the self-interested members they represent. Fortunately, there was an enabling environment, a small and closed working group. Negotiating in small working groups, particularly when it is conducted in a more professional manner, representatives are more likely to make decisions independent from their members (Kelman, 1981). As an informant who was a member of the legislative review group said, he could not recall that any members said, "Sorry, I cannot comment on that, I have to go to consult my members." He continued, "if they say 'sorry, we hear what you say, we have to take it back to the members,' that can delay for months, months, and months." The small working group actually served as a protector of those representatives from the intervention of their constituencies. A union check inspector in the review recalled that 95% of the judgments were made by three union inspectors, two of whom were on the review group. He noted that if the union has a policy on safety issues that the inspectors do not agree with, they can disregard that policy and then go back to explain what they believe they need to do.[32]

More important, these "right people" would have not reached any agreement about the new regulatory model if there were no webs of dialogue that brought them together. The web of dialogue is another important structural factor that facilitates the modeling process.

Webs of Dialogue

As discussed earlier, the actual outcomes in new mining statutes and regulations result partially from compromises between contending groups of model mongers. However, compromises are impossible to obtain if there are no webs of dialogue that bring contending groups of model mongers together to identify issues, create shared understandings, and resolve disagreements. Therefore, a web of dialogue is a structural basis for successful modeling.

Webs of dialogue can simply be informal interpersonal relationships. In this case, however, the webs of dialogue are institutionalized. They are exercised mainly through a series of tripartite initiatives. Some initiatives are associated with developing and implementing statutes and regulations such as legislation review groups and advisory councils. Other initiatives not directly related to legislative activities include annual health and safety conferences and some informal study groups.

Ayres and Braithwaite (1992, 57–58) define tripartism as one of three ways to foster participation. The first is to grant "the [public interest group] and all its members access to all the information that is available to the regulation." The second is to give "the [public interest group] a seat at the negotiating table with the firm and the agency when deals are done." The third is to grant "the [public interest group] the same standing to sue or prosecute under the regulatory statute as the regulator." Despite their discussion of tripartism mainly as "a strategy for implementing laws and regulations that have already been settled," they also acknowledge the role of tripartism in rule-making processes (Ayres and Braithwaite, 1992, 58).

In the rest of this section, several tripartite mechanisms and their influence on the diffusion of risk-based system regulation will be described.

Then, their institutional foundations, which explain why tripartite mechanisms can be formed in Australia, will be discussed.

Tripartite Mechanisms

Since previous sections have discussed tripartite legislative review working groups, this section starts with another important tripartite mechanism involving legislative activities: tripartite mine safety advisory councils. As the term implies, the advisory council is an advisory body to ministers who are in charge of mining safety in both NSW and Queensland. According to an informant, despite being an advisory body in name, "they have quite a lot of power to change legislation." In Queensland, the tripartite Coal and Mining Safety and Health Advisory Council was established under CMSHA 1999 and MQSHA 1999 as one important pillar of the laws. What distinguishes advisory councils in Queensland from those in NSW is the voting system used to make recommendations to the minister.

In NSW, provisions for the Mine Safety Advisory Council were added to the 1992 Mining Act in 2003. Section 341 of the Act states:

> (1) The Minister is to establish a Mine Safety Advisory Council that includes representation from peak industry and employee organizations.
> (2) The Mine Safety Advisory Council has the following functions:
> (a) Providing advice to the Minister on any policy matter relating to occupational health and safety in mines.
> (b) Any other advisory function relating to occupational health and safety in mines that is prescribed by the regulations.
> (3) The regulations may make provisions for or with respect to the constitution, members and procedures of the Mine Safety Advisory Council.

Following the recommendation of the 1997 *Mine Safety Review*, the first Mine Safety Advisory Council was convened in 1998 to implement the recommendations of this review. It was subsequently established in legislation (section 341 of the Mining Act) in 2003 (Wran and McClelland 2005 , 137). Apart from representatives from the government, unions, and employers, Section 47 of Mining Regulation 2003 states that the minister can appoint no more than two independent members, and the chairperson of the Council should be an independent person.[33]

One major purpose of the MSAC is to promote dialogue and information exchange among major stakeholders. As Wran and McClelland (2005 , 137) show, the purposes of the MSAC in NSW are to (1) improve safety across the mining industry in NSW, (2) provide strategic advice to the Minister on industry safety and health matters, and (3) facilitate consultation and exchange of information across industry sectors and among unions, employers, and the state government.

The advantages of tripartite working groups and advisory councils in developing and implementing legislation and regulations are many. First, it is helpful to get inputs from major stakeholders. Second and more important, this tripartite approach creates "ownership" of the legislation by involving all major parties. People tend to comply with rules that they develop themselves. As a government inspector explained, "when you come out with new legislation, the ownership is not government legislation. It is our legislation."[34] The active involvement of unions in developing and implementing legislation is considered as one of the reasons they at least agreed on the general direction.[35] Third, these tripartite mechanisms, bringing people from all parties together, can help to resolve disagreement. As Tony Maher, the general president of CFMEU, noted, the MSAC "provides a good forum to narrow the differences, if not achieve consensus" (Maher, 2006). An example is the one mentioned earlier, when developing recognized standards on the issues of drug and alcohol testing, the industry and unions could not agree on many issues for a long time. To build shared understandings, the government invited experts to the meetings of the tripartite advisory council to provide scientific knowledge on these issues. Eventually, disagreement on drug and alcohol testing has been resolved by passing a recognized standard. Similar stories happened in resolving disagreements on other issues such as fatigue management.

Such tripartite initiatives have also been widely used in other non-legislative activities. One example is annual health and safety conferences. The organizing committee includes representatives from all three parties. Since the organizer is tripartite, there was active participation by government inspectors, mining companies, and unions. Researchers also participated to present their research findings. According to informants, each conference usually included several hundred practitioners representing mining companies, managers, miners, crew leaders, union organizers, inspectors and regulators, politicians, and ministers. At the conferences, information, case studies, and research findings were shared. These conferences have taken place for at least 20 years. An informant considered these conferences as "the biggest factor" in explaining why Australians have a general agreement on the direction they should move toward. He also recalled that he saw the video of the Pipa Alpha disaster in the offshore industry and listened to a presentation about the results of the inquiry into that disaster. He said that it was a "really powerful message" for him.[36] These conferences provide wonderful opportunities for learning from others' experiences. According to social cognitive theory, on most occasions, people do not learn from direct experience but from vicarious experience by observing the consequence of others' activities.

A second example is informal tripartite study groups in Australia. For instance, in November 2003, the Department of Natural Resources, Mines and Energy in Queensland funded a tripartite mining safety study tour group that traveled to Canada, the U.S., and the U.K. to compare industry safety and health performance and regulatory frameworks with Queensland's. According to information in the Mine Inspectorate Safety Newsletter published by the Department of Natural Resources, Mines and Energy in Queensland, the study tour was proposed by the Minister for Natural Resources, Mines and Energy at the 2002 Queensland Mining Industry Health and Safety Conference, the tripartite initiative mentioned above. The tour group consisted of two persons from the industry [Kam Leung (Cannington mine) and Mitch Jakeman (Anglo Coal Australia Pty Ltd)], two from unions [Greg Dalliston (CFMEU) and Hag Harrison (AWU)], and one from government (Roger Billingham, a senior inspector in NRM&E) (DNRME, 2004).

According to an interviewee who was a member of the group, the study tour had two benefits. First, it "reinforced" the belief among all the parties that Queensland's new mining legislation and regulations based on risk management were on the right track. This is because, during visiting some mine sites, they had seen some obvious risks that were ignored in some countries that over-emphasize compliance with laws. Second, it increased the level of trust because people understood each other more through frequent conversations during the tour and better personal relationships.

Institutional Foundations

As previously stated, these webs of dialogue based on tripartite initiatives are institutionalized. In an interviewee's words, tripartism is a "long-standing approach." This is because tripartite mechanisms in Australia have their institutional foundations in industrial relations, industrial culture, and political systems.

Industrial Relations

Many European countries and Australia have had strong trade union movements. According to an informant, the strong union movement in Australia was partially because of the compulsory voting system, which required eligible voters to vote. If not, they could be subject to punishment such as fines and community services. The result was that the whole working class turned out to vote.[37] In addition, unions are closely connected to the Labor Party. As a result, a "power block" is formed and opposed to another powerful group—employers. An informant explained:

> If we deal with sectoral issues that are not labor-related, for example, we are dealing with the environment; trade unions would just be one group amongst many. You might expect major environmental groups to have a seat at the table. But if we talk about industrial relations and OHS, that would be fundamentally connected with what trade unions do. Trade unions would claim a monopoly on the rights on behalf of workers.

Owing to the strong union influence, legislation in Australia requires consultation processes to make sure unions have the input into legislation. "Consultation has always been there because it has to be there."[38] An interviewed inspector explained how absolute the requirement of consultation is. He said that the consultation requirement in Queensland has reached a "ridiculous stage." Unions must be involved in "very, very minor changes that won't have any impact" on how legislation works. Since consultation is an absolute requirement in the legislative process, a small tripartite working group itself becomes an especially appropriate consultation process. As an informant noted, "the best way to get your consultation is to have your people get together and work together."

Moreover, widespread tripartite initiatives result from not only strong unions' influence, but also increasingly harmonious industrial relations in recent decades. An informant said that he grew up in the era when unions in Australia were very adversarial. This industrial relation has changed "a lot" today. The reasons behind this shift are many, e.g., the change of culture in the society in general and increasing emphasis on corporate social responsibilities. Another important reason is the growing recognition of a common fate among employers and employees, as discussed earlier. Both employers and unions have realized that many situations are not win-or-lose. Something that makes mines profitable can also make miners wealthier. Consequently, unions today tend to exercise their "industrial muscle" differently by having more harmonious industrial relationships, particularly in the area of health and safety.[39] For example, the president of CFMEU called for more cooperation in an industry OHS conference.

> My message is simple, none of us can pick and choose. If we are going to take a cooperative approach then it has to be in all areas. It also has to be sustainable over a period of time. We shouldn't chop and change depending on the political color of the government of the day. . . . To deliver cooperation it will take very strong leadership and determination, but remember this—the alternative is Monty Python and the Holy Grail—and depending where we are in the commodities cycle, we'll take turns being the Black Knight.[40] (Maher, 2006, 10)

Industrial Culture

In addition to the institutional foundations based on industrial relations, there is a relatively more cooperative tradition in the Australian coal mining industry than in some other countries, such as the U.S. The Australian mining industry is a large industry in terms of capital and production, but it is more like a small community in the sense of limited geographic locations and culture. Observing from a background of metalliferous mining, an interviewee said:

> I find that coal miners are a very, very close-knit group of people; whether it's the miners, mine owners, or regulators, they are very, very close-knit, almost an incestuous relationship. They all know each other because it is a small industry. They seem love being told what to do. But metalliferous miners just want to get on with it. This is the culture difference between these two groups.

This closed culture has formed for a number of reasons. Some are general characteristics of the coal mining industry. Others are somewhat specific to Australia. Distinguished from metalliferous mining, coal extractions are long-life businesses. Some coal mines can last more than 100 years. This results in a fixed community where "father, son, grandson . . . all work in the same mine and all work together."[41] This is a common feature of the coal industry in Australia and other places around the world.

However, the next two Australia-specific characteristics make the industry's culture closer than that in many other countries. The first is geographic. According to an informant, coal fields in Australia are concentrated and also close to the coastal municipal areas. For example, the coal field in NSW is 100 miles north and 100 miles south of Sydney. Many miners who work in these coal mines live near the cities. This is different from metalliferious mines that are located in dispersed areas. Hence, people in the coal mining industry tend to have a more homogeneous culture because most of them come from the same region. Second, Australia has a relatively small inspectorate. Inspectors are assigned to a certain number of mines. They work with those mines for many years. An interviewee said:

> Even though we may have 20,000 people working in the coal mining industry in Queensland, it is amazing how many people and mines the chief inspectors work with. The miners, those operating longwalls, operating shuttle cars—they know him. Everyone knows everyone else. I don't know if it is good or bad. It is interesting.

It is not hard to imagine that, in such a small community in which people have a homogeneous culture and many know one another, people tend to cooperate and communicate with one another.

Political System

The third structural factor that "forces" the formation of webs of dialogue is the parliamentary political system. Under parliamentary political systems, the process of changing mining legislation is dominated by the executive, particularly people in small legislation and regulation review groups. The parliament's influence is very limited. When he talked about the role of the parliament, an inspector noted:

> Parliament itself has no real influence other than passing legislation and supports it when it was put forward. One of the reasons is that we work so hard with our tripartite groups . . . the influence is already from the industry and tripartite input is up to the stakeholders.[42]

He continued to tell an ironic story of the parliamentary debate about the new mining safety bill in Queensland:

> When it goes to the parliament, mining legislation has one of the longest debates we can recall. Believe it or not, it is almost totally supported by everybody, but everybody wants to have their say. I can recall only one amendment in parliament. They stuffed it up because it is all wrong. I was sitting in parliament when they were debating this proposed amendment. I could see they screwed it up. I wasn't allowed to say anything because only members of parliament can speak. It is funny because you have to go back next day to change it. It was a long debate because people started getting passionate about it, which is unusual. It went through as we wanted despite the 15 hour debate.[43]

As a result, interventions from politicians and lawyers are to a large extent prevented. The modeling process is controlled by "mining people and experienced people."[44]

Under this political system, formation of webs of dialogue is forced. This is because participants in small working groups know there are no allies in the legislature where they can turn for help. The only way for them to get what they want is to sit down at the table and negotiate seriously. If they leave the negotiation, they almost cannot get anything. An inspector talked about this:

> We don't have that sort of lobby groups in Queensland. If they lost conflict about resolution in our working groups, there is no doubt about it they will go to our minister. One group is always running to the minister. We never have lobby groups from outside going to parliament and influencing parliament on mine safety at all.

When people are forced to sit at a table surrounded by safety professionals with little interference from elected officials, more open and genuine dialogue is more likely to emerge. Self-interested political games are to a large extent prevented. As an informant stated:

> I can show you some of the discussions we have in the meetings. They are quite forthright. In Australia you can get together, you can argue and you can talk openly to each other. Say what you think. . . . Some people always blame workers and some people always blame the management. But generally, occupational health and safety is managed on the issues, not on the conflicts.

CONCLUSION

In sum, the success of moving toward a risk-based system regulatory model in Australian coal mining safety regulation is first because of the imitative pressure from general OHS regulation, which set the direction and framework for debate. However, even though imitative pressure is in place, changes may not occur if no policy window opens and no one champions the change. These two conditions were met in this case. The change of government in NSW and Queensland opened the policy window. When the window opened, there were model mongers (senior government inspectors, mine operators and managers, and union check inspectors) who drove the change. They had the cognitive capacity to understand the value of the risk-based system regulatory model and also perceived anticipated benefits in this model.

Nevertheless, even though there are people who drive change, the modeling process usually involves conflicts between different groups of model mongers because of their different interpretations of the model based on their interests. To model through, there must be someone who helps different parties to build shared understandings. In Australia, model missionaries and mercenaries played an important role. They had many opportunities to shape people's cognition through such activities as providing training and participating in accident inquiries and mine safety reviews.

Moreover, different parties also need platforms that bring them together to create shared understandings and make compromises. Such platforms exist, too, based on webs of dialogue. These webs are not dispersed involving many actors, but very concentrated, intensive, and closed webs dominated by three major stakeholders. The formation of this type of webs is the result of industrial relations, a cooperative culture, and little interference from other political institutions, such as the parliament. The closed and tripartite webs of dialogue in turn prevent the influence of self-interested politics and encourage more open and genuine dialogue. This echoes what Braithwaite (1993, 92) calls for—a "policy space where mutually respecting interest groups really talk to each other about their concerns." He also indicates there is more opportunity in Australia to develop such "policy space" than in other countries.

> In Australia, we have a long way to go before reaching such a pass. On the other hand, there is much more of the makings of such a constructive regulatory culture in Australia than in many other countries. There exists in Australian regulatory communities a kernel of mutual respect and fair play that can be nurtured. (Braithwaite, 1993, 92)

In addition, the research again suggests that Braithwaite and Drahos's (2000) framework needs to be modified. As with the findings in the two previous chapters, those in this chapter indicate that the reactive sequence of regulatory change in their framework is actually a combination of both reactive and proactive forces. As this case shows, governments not only react in response to disasters; they also act proactively when the control of government changes. Moreover, model mongers do not always have multiple models to propose. They may only have one model when imitative pressure is strong.

Having identified the factors that explain the successful modeling of risk-based system regulation in Australia, the next chapter will examine the factors that prevent the U.S. government from fully embracing the model based on a comparison with the findings in this chapter.

6 Barriers to the U.S. Government's Move toward Risk-Based System Regulation

INTRODUCTION

As shown in the last chapter, Australian coal mining safety regulation has embraced a risk-based system regulation. By contrast, the U.S. coal mining industry still operates under a highly prescriptive regulatory regime. This does not mean, however, that no efforts have been taken to move away from the prescriptive regulation in the U.S. In fact, between 2001 and 2004, the national regulator MSHA put forward several policy initiatives that at least to some extent moved toward the risk-based system regulatory model. The purpose of these new initiatives was to encourage and help operators to establish management systems and build up their capacities to proactively identify, analyze, and control hazards. Nevertheless, MSHA's effort was not as successful as their Australian counterparts. This chapter seeks to answer this question: what are the barriers keeping the U.S. government from moving toward risk-based system regulation?

The previous chapter showed that three variables explain why the risk-based system regulatory model was adopted in Australian coal mining safety regulation: strong imitative pressures, the influence of the "right people" (model mongers, missionaries, and mercenaries), and the existence of webs of dialogue. The research finds that these three factors also explain why the U.S. government has not successfully moved in that direction. First, there is no strong imitative pressure from the general OHS regulation in the U.S. OSHA regulation is at most an intermediate form of command and control regulation and risk-based system regulation. Second, although there are also model mongers, missionaries, and mercenaries, they are all in weak positions. Model mongers in MSHA encountered strong resistance from both inside and outside the agency. For mining operators and managers, some negative perceptions of the risk-based system regulatory model, such as liability concerns, prevent them from becoming strong model mongers. Moreover, unlike Australia, there are no model mongers in unions who can bring the unions' position closer to that of the industry and government. Model missionaries and mercenaries in the U.S. lack the channels that their Australian counterparts have to exercise their influence. In addition, even if model mongers, missionaries, and mercenaries had stronger influence, it would be still difficult for the U.S. to move toward a risk-based system regulatory model because there are few webs of dialogue, such as the tripartite mechanisms in Australia, to bring different parties together. The lack of tripartite mechanisms in the U.S. mainly results from an adversarial industrial relation and culture, as well as from fragmented political institutions that open too many avenues for the industry and unions to pursue their interests.

The rest of the chapter will discuss the influence of the three factors in greater detail. Before examining these factors, the existing coal mining safety regulatory framework in the U.S. will be briefly introduced, followed by a description of the MSHA's effort to move toward the risk-based regulatory model.

EXISTING PRESCRIPTIVE REGULATORY REGIME

U.S. coal mining safety regulation relies heavily on the prescriptive approach, which is characterized by strict standards and tough enforcement and sanctions. Almost every federal statute was enacted as a response to political and social pressure resulting from mine disasters. According to Grayson and Watzman (2001), legislation before 1969 did not give regulatory agencies adequate authority to establish and enforce safety standards.[1] In response to the Farmington explosion in West Virginia in 1968, Congress passed the Coal Mine Health and Safety Act (the Coal Act) of 1969, which became the first law with teeth (Gunningham, 1999). It required four annual inspections for underground coal mines and two annual inspections for surface mines, authorized miners to request federal inspections, and imposed mandatory fines for all violations. In 1977, the Congress enacted the Federal Mine Safety and Health Amendment Act (the Mine Act), which further increased the enforcement power of federal inspectors and sanctions against mine operators who fail to comply with the laws (*ibid.*). The 1977 Mine Act set up the foundation of the current regulatory regime.

According to Grayson and Watzman (2001, 5), the major provisions of the 1977 Act are as follows:

- The enforcement agency was moved from the Department of Interior to the Department of Labor and renamed the Mine Safety and Health Administration.
- Specific standards for designing and operating coal mines were established, from roof control to ventilation of mining sections.
- Four inspections of underground mines and two inspections of surface mines were required annually.
- Mine operators were required to conduct pre-shift and on-shift inspections.

- The involvement of miners and their representatives in safety activities was increased.
- Violation of safety standards was subject to penalties of up to $55,000.

In 2006, some provisions of the 1977 Mine Act were amended by the Mine Improvement and New Emergency Response Act (the Miner Act), which was passed by Congress in response to the Sago disaster that occurred on January 2, 2006. As Chapter 4 discussed, the Miner Act requires underground coal mine operators to:

- notify the Secretary of MSHA within 15 minutes after the occurrence of a fatality or accidents that have the potential to cause death;
- develop and adopt a written accident response plan, which provides post-accident two-way communications, post-accident tracking, and a wireless two-way communication system and electronic tracking system within 3 years;
- make available two experienced rescue teams that are able to respond in 1 hour.

The Miner Act also increased the amount of civil penalties to a minimum of $2,000 for any citation and a maximum of $220,000 for flagrant violations under section 104 (d) (1) and (2). As a result, the current coal mining safety regulations in the U.S. have become even more prescriptive.

In addition to federal statutes, MSHA also has issued regulations, including requirements on testing and approval of mining products, education and training, civil penalties for violations, as well as mandatory health and safety standards. Unlike Australia where coal mining safety is a matter only at the state level, both the federal and state governments in the U.S. have authority for regulating coal mining safety. Besides federal statutes and regulations, many states also have additional statutes and regulations. However, some states simply follow the federal statutes and regulations.

EFFORT TO MOVE TOWARD RISK-BASED SYSTEM REGULATION

Although there has been no change in existing statutes or regulations, the national regulator (MSHA) has put forward some policy initiatives that at least move toward the risk-based system regulatory model. Most of these initiatives were put in place during 2001–2004 when David Lauriski was an assistant secretary of the Department of Labor, the head of the MSHA.

As Braithwaite and Drahos (2000, 561–562) suggest in their reactive sequence of regulatory change, the most important actors driving regulatory change are not strategic individual entrepreneurs from business or NGOs, but key individual entrepreneurs who have direct organizational control of the state's capacity to act in response to disasters. However, based on my observations in this study, individual entrepreneurs within the government who advocate regulatory innovation can initiate regulatory changes to some extent without the occurrence of major disasters. Lauriski is such an entrepreneur who drove MSHA's policies to move from purely enforcement toward the direction involving more risk management.

Lauriski is actually a model monger of the risk-based system regulatory model. According to social cognitive theory, whether a new practice can be adopted by people to a large extent depends on their cognitive capacity, which is formed mainly based on their previous experiences. As the last chapter discussed, the risk-based system regulation makes sense to the senior inspectors in the Australian government because they have experience as mine managers. Lauriski has similar experience. Before he was appointed assistant secretary, he had more than 30 years of experience in the industry. He is a certified safety professional, and he worked as a mine safety director and manager in mining companies. He also brought in two deputy assistant secretaries who had more than 25 years of experience as safety professionals and managers. The experience in mine safety management gave Lauriski and his colleagues the cognitive capacity to fully understand the limitations of the prescriptive regulatory approach and the value of risk management.

When Lauriski took over the MSHA, he and his colleagues recognized major weaknesses in the prescriptive regulatory regime in the U.S. The first is that the prescriptive regulation had created a hostile relationship between the industry and MSHA. Second, almost all regulations and enforcement focus on physical conditions that can be engineered, such as issues related to roof control and ventilation. There is almost no language in regulations that deal with human behaviors and the safety management processes. Therefore, Lauriski and his deputies believed that people's behaviors and management systems should be significant components of accident prevention. As an interviewee said:

> Here is the thing, you can conduct physical hazard inspections, you may find hazards, and that is great. But if their management systems do not reinforce certain aspects of the workplace, that physical hazard will come right back. It will be there until another inspector finds it. But if they focus on their safety management systems, their own self-inspections, for example they have their own miners trained to find their own hazards, MSHA comes in to do an audit to make sure their system are not just sitting on the shelf. To me that makes much more sense.

Based on this recognition, Lauriski and his colleagues had different understandings from many people about the role that MSHA should play. Many who were in MSHA, Congress, and unions believe that, according to the 1977 Mine Act, the role of MSHA is a pure enforcement agency that enforces existing regulations. However, Lauriski and his colleagues' interpretation was that the Act authorized MSHA to find ways to reduce injuries, illnesses, and deaths in mines. To achieve this goal, one component is enforcement. But there are "other components including education, technical support, and anything that would move upward to better performance."[2]

Therefore, their philosophy was that MSHA, along with enforcing the law, has obligations to help miners and mining operators to better understand how to identify and mitigate risks. During an interview, Lauriski said that he spent about a year traveling to every district office of MSHA. The message he sent to the employees was as follows:

> Look, we have more to offer than just writing a mine operator a violation of law. We have the ability to influence what happens in mine safety because mine safety is more than just regulation. It is about behaviors; it is about getting people to understand how to identify hazards . . . how to convince them to make safety a value rather than just to do what they feel convenient.

Based on this philosophy, MSHA, led by Lauriski, adopted or attempted new policy initiatives that placed greater emphasis on cooperative strategies in addition to enforcing the law. These initiatives addressed the areas of compliance assistance, enforcement, training and education, accident investigation, and stakeholder relationships.

Compliance Assistance

MSHA under Lauriski emphasized a "balanced approach of using enforcement, education and training, and technical assistance," which was called the "Triangle of Success" (Lauriski, 2004). A key component of this strategy was compliance assistance that was carried out in three ways. First, inspectors were expected to do more than inspections of the physical workplace during mine visits. They were also expected to "assist mine operators to determine the root causes of hazards that lead to both violations and incidents" (*ibid.*). As an informant explained:

> One of the things that we emphasize to our employees is "look, when you enter a mining operation, if you see a violation, first of all we expect you to issue a citation, but when you issue a violation, we also expect you to have a conversation with the mining operator about the importance of that violation and not allow that to reoccur from a safety perspective."

Compliance assistance opens up channels of communication, which allow inspectors to share their experiences and knowledge with miners and operators. Second, MSHA engaged operators and miners in analyzing fatalities, including both causes and prevention, and discussing best practices through national webcasts. The purpose of the webcasts was to bring all stakeholders into a dialogue (Lauriski, 2004). The third way to provide compliance assistance was through a small mine office, which opened in 2003. The function of the office is to help small operators, which have difficulties in complying with regulations, to develop basic safety processes, to understand how to comply with regulations, and to provide assistance with their education and training (*ibid.*).

Training and Education

The expectation that inspectors conduct compliance assistance during inspections requires different skills from "going by the book" and writing citations. According to an informant, during the time when Lauriski was in office, MSHA tried to make inspectors "not just paper writers, but to be paper writers and thinkers." To achieve this goal, the National Mine Health and Safety Academy started to offer more training for inspectors and the industry on behavior-based safety and hazard identification. An example is the SLAM process introduced by MSHA as an education and training initiative, as mentioned in Chapter 4.

SLAM has been widely accepted by the industry because it integrates well with the BBS process that many American mining companies have developed. Blacksburg Coal is an example. When a safety manager talked about how SLAM contributes to its BBS program, he said:

> In 20 plus years of mining, MSHA has come out with a lot of programs, and that SLAM program is the best that I have seen. That couples with our BBS program so well. As a matter of fact, since MSHA put out the SLAM program, we have used SLAM in all our annual retraining and our BBS training. You stop, look, analyze, manage; that is where our whole safety process started at that point. MSHA just came in with that same process that just marries up so well [with ours]. We have integrated the MSHA SLAM process into our process . . . SLAM is much easier for people to understand. It just gives us a better definition to labor about our process.

Accident Investigation

To achieve the same purpose as compliance assistance—helping mine operators and miners to identify and mitigate risks—MSHA required all investigators across the country to do a root cause analysis in their accident investigations. According to an informant, traditional investigation had stopped as soon as the agency found a fault. This approach would "never be able to really determine what the causal factors that caused the accident were."[3] I checked accident reports in the MSHA digital library. There is only a general section called "accident discussion" in all accident reports before 2003. Since 2003, all accident reports have included an independent section on root cause analysis.[4]

Stakeholder Relationships

MSHA under Lauriski adopted more cooperative strategies with major stakeholders, rather than relying purely on deterrence. The first major cooperative strategy involved stakeholders' meetings, starting in 2002. According to an interviewee, MSHA held about 150 stakeholders' meetings in 2002 alone. The issues at these stakeholders meetings ranged from discussion of the strategic objectives of the industry to reviewing rule-making.

A second important cooperative strategy is MSHA's alliance program, which started around 2003. Learning from a similar program in OSHA, MSHA's alliance program focused on signing agreements with trade associations or unions that commit to getting their memberships to agree to reduce fatalities and injuries. According to information on the MSHA website, MSHA has had ten national alliances.[5] Eight out of these ten alliance agreements were signed during 2003 and 2004 when Lauriski was the head of MSHA.

Enforcement

In addition to the policy initiatives that were put in place, there were also some attempts to streamline the inspection process, which allowed MSHA to allocate its enforcement resources more efficiently and effectively. According to ICF (2003), after Lauriski was appointed, MSHA asked ICF Consulting to evaluate how to improve the inspection program. Part of the evaluation was a reexamination of the recommendations proposed by two internal work groups, the Coal Reinvention Work Group and the Metal and Non-Metal's (MNM) Voluntary Compliance Partnership (VCP), in the late 1990s.[6] ICF Consulting recommended that MSHA implement these recommendations, including the MNM's VCP initiative. The VCP initiative, which is similar to OSHA's Voluntary Protection Partnership (VPP), was designed to recognize those mines with excellent safety and health performance by offering them the status of being inspected less frequently than mandated levels. According to the MNM VCP proposal, approximately 5 percent of total operations would be eligible for VCP status. These operations must have written safety and health management systems, excellent training on hazard identification and accident prevention, and internal audit systems. According to an MSHA employee, ICF's evaluation is the only government report that he is aware of that looked at risk-based regulation.[7]

The focus of MSHA's efforts during Lauriski's tenure was thus shifting from a purely compliance-driven mentality to a mentality involving more proactive accident prevention. The effort was aimed at increasing operators and miners' cognitive capacity to proactively identify and control workplace hazards. To achieve this goal, MSHA had to first enhance the cognitive capacity of its inspectors. That is why it adopted initiatives such as SLAM, behavior-based safety, and root cause analysis. As Lauriski said in his remarks to the Indiana Mineral Aggregates Association,

> MSHA is committed to helping you and the Indiana mining industry bring down even further your injury and fatality rates. And we are doing it by changing our structure, philosophy, and attitude. (Lauriski, 2004)

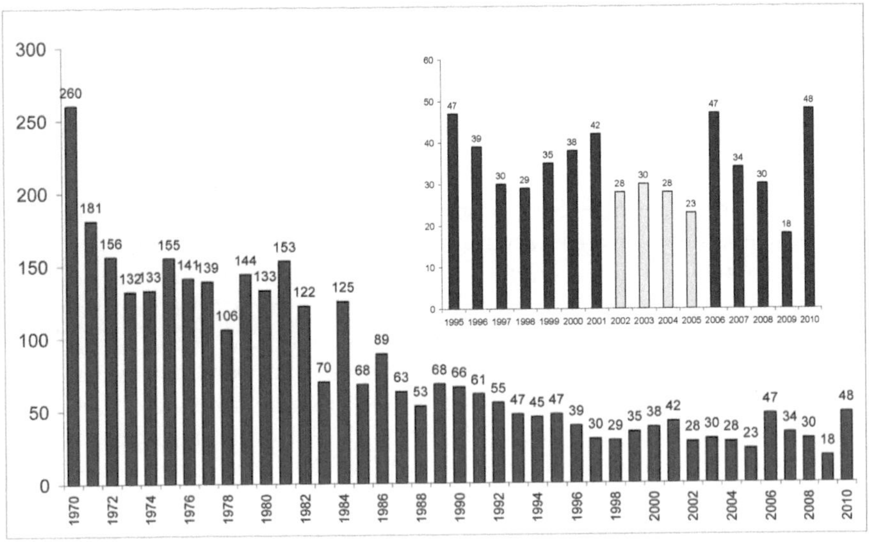

Figure 6.1 Annual coal mining fatalities, 1970–2010.
Source: MSHA (2011).

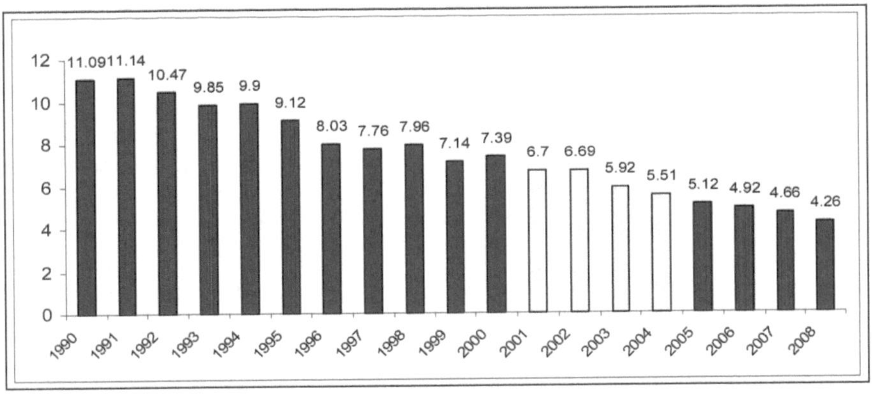

Figure 6.2 Total incidence rate in the coal mining industry after 1990.
Source: MSHA (2010c).

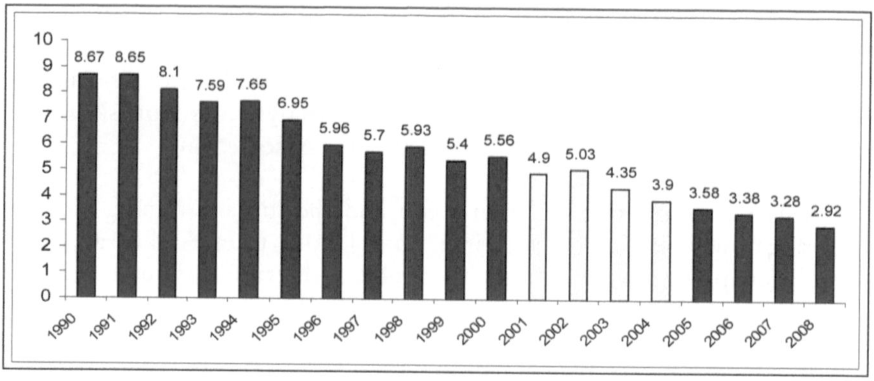

Figure 6.3 Nonfatal days lost incidence rate in the coal mining industry after 1990.
Source: MSHA (2010c)

As a result, the role of MSHA under Lauriski moved away from being a pure police officer toward serving as a cooperator and facilitator of the industry to achieve better safety performance. As an interviewee said, "if we work as a community not a prosecutor, we have a better chance to improve safety in this country than we did by strictly being an enforcement agency." Safety performance in the mining industry improved significantly during Lauriski's tenure. The fatality numbers between 2001 and 2005 were the lowest in history (Figure 6.1). All other statistics, such as total incidence rate and nonfatal days lost (NFDL), also continued to move downward (Figure 6.2 and 6.3).

However, these MSHA's policy initiatives were at best only a partial success because of strong resistance from both inside and outside the agency. In the words of an informant, many of those initiatives did not "really get off the ground." For example, the MNM's VCP initiative was never implemented. This is partly because there is little room for the MSHA to adopt a program like VCP within the legal context of the 1977 Mine Act. Unlike the OSHA Act, the Mine Act mandates four annual inspections for underground mines and two inspections for surface mines. Moreover, VCP was also resisted by unions because they considered it as something not consistent with the 1977 Mine Act.[8] When an informant talked about the VCP initiative during the interview, he said:

> We had actually developed the actual process. UMWA got wind of that; they went to Congress and effectively caused us to halt our work on that. That never came back.

Moreover, despite being encouraged, compliance assistance has never been "fully institutionalized and incorporated into the inspection process" (ICF, 2003, 2). According to informants and ICF (2003), compliance assistance activities vary among inspectors. In other words, some inspectors provide compliance assistance while others do not. Moreover, the *Coal Inspection Procedures Handbook* defines very specific steps that an inspector has to follow. There is no flexibility that allows them to conduct compliance assistance.[9]

More important, Lauriski resigned in November 2004, partly because of personal reasons and partly because of "miserable battle fighting everyday on the political front."[10] Most of the policy initiatives remained in place after he left office, such as stakeholder meetings, root cause analysis, and the alliance program, but MSHA's move toward the risk-based regulatory approach stopped. Moreover, after the Sago disaster occurred in 2006, MSHA to a large degree shifted back to a purely prescriptive approach. As Figure 6.4 and 6.5 show, the number of citations and violations per inspection hour increased dramatically after 2006.

Lauriski's contributions were recognized by several interviewees from both the MSHA and the industry. As an informant said, "had David Lauriski been in office, had Sago not happened, we probably would have some kind of risk management." The question is why these innovative efforts initiated by model mongers including Lauriski and his colleagues between 2001 and 2004 were not as successful as those of their Australian counterparts in terms of moving the regulatory framework away from a prescriptive approach toward the risk-based system regulation. This question will be addressed in the next section.

BARRIERS TO MOVING TOWARD THE RISK-BASED SYSTEM REGULATION

As the findings in the last chapter suggested, three major factors led to the success of regulatory reform in Australia: strong imitative pressure from general OHS regulation; strong influence of model mongers, missionaries, and mercenaries; and the existence of webs of dialogue. This section will examine these factors to see what their influence has been in the U.S.

Imitative Pressure

As shown earlier, Australian coal mining safety regulation successfully shifted from a prescriptive approach to a risk-based system regulation partly because of the strong imitative pressure from the general OHS regulation. When governments in Australia considered a change in mining legislation, the direction was already clear, because the whole regulatory framework in the general OHS area had already moved to a risk-based system regulatory approach. The change of general OHS regulation in Australia was a result of the strong influence of the Robens model of regulation in the U.K.

Compared with the Australian case, there was no such strong imitative pressure toward a risk-based system regulatory model from general OHS regulation in the U.S. When the Robens Report came out in 1972 in the U.K., both the general OHS and the mining-specific regulations in the U.S. had already started on a different path. Congress had enacted the Coal Act of 1969 and the Occupational Safety and Health Act (the OSHA Act) of 1970. Unlike the Robens Report that advocated a more self-regulatory approach, both the Coal Act of 1969 and the OSHA Act of 1970 created a highly prescriptive regime that relied on strict standards and tough enforcement.

In a comparative historical study of the development of OHS regulation in Australia and the U.S., Clark (1999) reveals why the two countries chose diverse paths since the early of 1970s despite a similar beginning. According to her study, OHS regulations in both countries relied on state-based prescriptive standards and a "persuasive" enforcement style before the 1970s. During my interviews, an informant commented that coal mining safety regulation in the U.S. before the late 1960s was similar to that in Australia, which relied more on self-compliance.[11] However, after the Coal Act of 1969 and the OSHA Act of 1970, both coal mining safety and general OHS regulations in the U.S. started to depart from previous practice by establishing a more centralized, rule-oriented, adversarial, and punitive regulatory approach. By contrast, regulatory reform in Australia was much less revolutionary and moved continuously along the path of a cooperative regulatory approach (Clark, 1999, 98).

Clark (1999, 94) explains that the regulatory divergence in these two countries resulted from the distinct configurations of political and social institutions and the different constellations of political actors involved in regulatory reform. The fragmented nature of political institutions in the U.S. was more open to public interest groups than the parliamentary system in Australia. The coalition advocating OHS regulatory reform included a wide range of actors, not only organized labor, but also Ralph Nader and public interest groups. The labor movement was also linked to the environmental and consumer movements. The linkages of different movements had influence on the perceived "problem" and its "solutions." Consequently, the "problem" was defined as the capture of government by industries. As such, "cooperation, discretion and flexibility were designed out, while adversariness and strict adherence to rules and procedures were designed in" (Clark, 1999, 99). Kagan (1991) also indicates that the regulatory style in the U.S. became increasingly adversarial after the 1960s. This is because there was a mismatch: on the one hand, the movement leaders demanded that government do more in regulating workplace safety and health and in providing environmental and consumer protection; on the other hand, they were suspicious of government authority and demanded further fragmentation and constraints on government.

As social cognitive theory suggests, when an adversarial style of regulation became the primary cognitive framework in the U.S., the chances of attention being paid to new models such as the 1972 Robens Report were significantly reduced. Added to the weaker historical connection with the

U.K. in the U.S. than was the case in Australia, the influence of the Robens Report in the U.S. was very limited compared to its "immediate and direct" impact in Australia.[12] Little evidence of serious discussion about the Robens model of regulation in the U.S. can be identified. In addition, none of the informants, including experienced safety professionals in the industry and MSHA officials, knew much about the Robens Report.

Since the late 1970s, however, general OHS regulation in the U.S. has gradually embraced more flexible and generic standards, partly because of the influence of neo-liberal ideology and partly because of OSHA's resource shortfalls (Needleman, 2000). Some examples of the more flexible regulatory approach are California's Cooperative Compliance Program (CCP), and OSHA's Voluntary Protection Program (VPP) and national CCP program.[13] These initiatives, however, were designed to encourage but not mandate health and safety management systems. The general regulatory framework set up by the OSHA Act of 1970 was still prescriptive and adversarial. The new initiatives that placed greater emphasis on cooperation and flexibility were only deviations from the main path. Therefore, general OHS regulation in the U.S. after the 1990s was at most an intermediate form between the Robens model of regulation in Australia and the OSHA command and control regulation of the 1970s.

Moreover, even these small departures from command and control regulation moved painfully owing to the strong suspicions of both labor and businesses. For example, organized labor considered OSHA's national CCP program a retreat from "OSHA's historically pro-labor stance" (Needleman, 2000, 79). Even the business community that constantly expressed its wish to have more cooperative regulatory approaches did not favor some initiatives either. As mentioned in Chapter 4, OSHA's national CCP program was struck down by the court in 1999 because of the challenges from some employer associations.

This was the imitative pressure that MSHA faced when it considered moving away from a prescriptive approach under Lauriski's leadership. Actually, many of the adopted or attempted MSHA initiatives between 2001 and 2004 were imitations of OSHA's programs, including compliance assistance, the alliance program, and the VCP initiative. Unlike its Australian counterpart, what MSHA imitated was a hybrid approach of the Robens model of regulation and traditional command and control regulation. As a result, MSHA's regulatory reform could at best go as far as what OSHA did. As a matter of fact, coal mining safety regulation in the U.S. is still far more prescriptive than general OHS regulation. For example, MSHA's VCP initiative, which attempted to recognize enterprises having exemplary safety records by reducing routine inspections, died even in the planning stage. By contrast, a similar program adopted by OSHA (VPP) in 1982 has worked quite well up to the present. MSHA cannot move as far as OSHA did toward more cooperative and systems-based regulation partly because the labor-management relationship is more hostile in the coal mining industry. As already mentioned, the MSHA's VCP initiative did not take effect because of the strong resistance of the UMWA.

Model Mongers, Missionaries, and Mercenaries

As the last chapter showed, risk-based system regulation could be modeled through in Australia because it has the "right people." These individuals include model mongers, such as senior government inspectors, mine operators and managers, and union check inspectors, who have the cognitive capacity to understand the value of the model; and model missionaries and mercenaries, such as Jim Joy, who distribute knowledge associated with the model and help different parties to achieve shared understandings. As previously stated, MSHA's move toward the risk-based system regulatory model has not been successful. One may ask: "is this because there are no right people in the U.S.?" The answer is no. There are also model mongers, missionaries, and mercenaries in the U.S. However, they are all in relatively weak positions. Comparing them with their Australian counterparts, this section examines the role of these actors in the process of MSHA's moving away from prescriptive regulation.

Model Mongers

The model mongers in Australia include three types of actors: senior government inspectors, mine operators and managers, as well as some union officials. This sub-section will compare these three types of actors in the U.S. with their Australian counterparts.

Government Inspectors

In Australia, senior government inspectors are the most important model mongers. They drive the shift toward the risk-based system regulatory model because they have the cognitive capacity to understand the value of risk management because of their experience as mine managers. In addition, they perceived the anticipated benefits of the new regulatory model—transferring responsibilities and completing regulatory jobs more cost-effectively.

The most important model mongers in the U.S. government were Lauriski and his colleagues. They were not inspector themselves, but they provided leadership to inspectors. Like senior government inspectors in Australia, they also had experience in the management of mining companies and therefore also understood the value of risk management. In addition, they saw the anticipated benefits of the model, at least as a more cost-effective way to allocate regulatory resources. That is why they brought in ICF Consulting to evaluate MSHA's inspection program and to find ways to better allocate regulatory resources. When Lauriski talked about the lack of inspectors in MSHA and how to resolve the problem by adopting risk-based regulation, he said:

> One important thing today is they don't have enough inspectors. Well, that is not the issue. First of all it is how you manage the resources. You manage the human resources; you get the work done. If you do it correctly, you don't need to have a group out there. I know mines today; they have 15 inspectors a day doing inspection. That is absolutely an inefficient way to manage that human resource . . . if you use them in the right way, not just from an enforcement perspective, who knows what kind of performance they could achieve if they got away from the idea of strict enforcement activities. Make enforcement one of the components of an effective way to improve safety and health in mines. But our politics ignore that.

Such model mongers in MSHA, however, were in a very weak position owing to resistance coming from both inside and outside the agency. The resistance from the inside mainly stemmed from MSHA employees who were "entrenched in the way they thought about things."[14] During the interview, an informant considered the difficulty of changing traditional bureaucratic thinking as one of three major barriers to change (in addition to the influence of Congress and the resistance of unions).

The resistance within MSHA can be attributed to a variety of factors. Some resistance is simply related to personality. Some people are open-minded, whereas others find it more difficult to accept change. Moreover, civil servants in government regulatory agencies, according to an interviewee, are more resistant to change because they have new political appointees every four years who may pull them in different directions.

> Government agencies are kind of a unique animal. Every four years, you get a new boss, and every new boss who comes to the agency has certain ways they want to do things. Someone is getting tired of it. You can understand why they are like that. . . . I have heard people say that "well, we will go ahead and do this, but we know that when Lauriski leaves, the agency are not going to have to do it."[15]

More important to this research, however, resistance within the MSHA to some extent results from the lack of cognitive capacity of inspectors to understand the value of risk management. A mine manager certificate is a prerequisite for being an inspector in Australia, but there is no such requirement to be a MSHA inspector. As a matter of fact, the threshold to be a MSHA mine inspector is quite low. The Mine Act requires that a mine inspector must have five years "mining experience." However, according to ICF (2003) and some informants, the five-year experience requirement is not necessarily in-mine experience. Education in a mining area can offset the five-year requirement. Therefore an individual holding an undergraduate degree in mining engineering plus one year of relevant apprenticeship training can meet this requirement.

According to the information on the website of the U.S. Office of Personnel Management, qualifications for a GS-7 level mine health and safety inspector (the lowest level) are as follows:

- Education: one full academic year of graduate level education
- Specialized experience: On-site safety/health inspection, analysis, monitoring, or evaluation work in occupational health, or mining, or closely related industries, such as building construction, excavation, and heavy industrial plant settings (OPM, 2009).

It can be seen that specialized experience in inspection is even not necessary in mining. Moreover, the experience base of inspectors is becoming weaker given the pressure for more enforcement that has been put on MSHA by Congress after the Sago disaster, and the fact that about 40% of MSHA's inspectors will be eligible for retirement by 2012 (GAO, 2007). According to some informants, MSHA has started to recruit inspectors with even less mining experience than before to meet the workforce needs. GAO (2007) also reported that the Department of Labor interpreted the statutory requirement of five years of experience as only a preferred but not an absolute requirement. This is one of the biggest problems that worry many respondents. One said that hiring inexperienced inspectors is totally wrong. Another interviewee argued:

> I will give you an analogy. When you get so many new people in, and the percentage of people who really don't have experience is growing, be aware of the China sector; they don't have experienced miners and inspectors. Oftentimes in China, they hire people who don't know much of mining.

The lack of cognitive capacity of MSHA's inspectors not only results from their short mining experience but also from their narrow scope of experience. Unlike Australian inspectors who have experience in all aspects of mining safety owing to their backgrounds as mine managers, some MSHA inspectors are discipline-specific. In the U.S., there are general mine inspectors as well as inspectors in special areas such as ventilation, roof control, and electricity. This difference was observed by an Australian expert on risk management.

> In Australia, to be a regulator you have to be a mine manager. The whole regulatory person [in Australia] is always different from the U.S. where people are trained in niches. . . . I could have this too simple. But people [in the U.S.] may have experience in mining, but not particularly as mine managers. Regulators [in Australia] are selected very

> carefully; the regulators are normally very experienced mining managers. . . . Some people think that approach is a higher quality regulatory approach. . . . People who are trained in ventilation only know ventilation and write tickets.[16]

The whole MSHA inspector system is thus designed to meet the requirements of a prescriptive regulatory regime in which inspectors are simply police officers who "go by the book" and find faults. They are not expected to have the competence to coach people at mine sites on managing their risks. Owing to their inadequate cognitive capacities, MSHA inspectors find it difficult to accept the change to risk-based system regulation, which requires that inspectors to have more comprehensive experience in mining safety management.

In addition to the internal resistance to moving toward a risk-based system regulation, MSHA also encountered external resistance coming from unions, the Congress, and the media. According to Bandura (1986, 159), when sociocultural change encounters privileged groups that have a vested interest in preserving existing social arrangement, "little change will result unless adopters are protected from unauthorized coercion." The model mongers in Australia had such protection. They worked in small working groups, which allowed them to make decisions independent of the members they represented. There was also little intervention from the parliament because the executive had gradually come to dominate the legislative branch in parliamentary systems. As a result, changes in mining legislation and regulations are controlled by "mining people and experienced people," but not legislators.[17]

Lauriski and his colleagues did not have such protections because of the deep suspicion of government authority and the more fragmented political institutions in the U.S. According to informants, Lauriski's biggest political disadvantage was that he was an executive of a mining company before he went to MSHA. Owing to this background and the very partisan Bush administration, any of his efforts to change from traditional prescriptive regulation were subject to tremendous suspicion and scrutiny from unions, some in the Congress, and the media. As a matter of fact, he was criticized as being "in the pocket of the industry."[18] This strong reaction was totally beyond his expectations. In a media interview, he said:

> I was a little naive [about dealing with the media and the overall level of scrutiny]. I expected it, but not to the level that I've seen. [The scrutiny] is largely because we've undertaken a lot of changes and sometimes change creates different perceptions. That's been a surprise to me. (Markley, 2004)

The "different perceptions" that he referred to mainly came from a contending group of model mongers with vested interests in preserving prescriptive regulations. Based on information collected from interviews and from media reports, a major criticism was that some MSHA initiatives, which were put forward during Lauriski's tenure, moved away from the intent of the Mine Act. The Act, according to these critics, mandated nothing but promulgating and enforcing regulations. For example, a summary of the testimony of Cecil E. Roberts, the President of UMWA, before the Senate in 2007 indicates this.

> Roberts noted for years, MSHA has subverted the will of Congress by promulgating regulations, policies, and petitions for modification that allow coal operators to take steps that are in direct contradiction to the 1969 Coal Act and the 1977 Mine Act. . . . Nowhere in that Act does it say MSHA needs to help the operators improve production. But many times over the past years, MSHA has been an ally in the operators' efforts to do just that. . . . "This trend has become much worse since 2001," Roberts said, citing decisions made by upper MSHA management to drop proposed regulations and instead focus on compliance assistance and collaboration with mine operators on regulatory initiatives. "The law isn't written to protect mine operators or their profits," Roberts said. "The law is written to protect coal miners' health and safety, period." (UMWA, 2007)

More specifically, some opponents claimed that MSHA under Lauriski spent more time on compliance assistance and less time on inspections than it had under previous directors. However, according to the testimony of David Dye, the acting assistant secretary of the Department of Labor before the Subcommittee on Labor, Health and Human Services and Education of the U.S. House, the level of inspections during Lauriski's tenure did not drop.

> Our completed inspection rates continue to rise. Our inspectors continued to complete 99 percent of the required inspections in coal mines between FY 2000 and FY 2004. During the same time period, the completion rate jumped from 74 percent to 90 percent of required inspections in metal and non-metal mines. (Dye, 2005)

Other opponents contended that, even though the inspection completion rate increased, the number of serious violations, fines, and criminal prosecutions dramatically declined compared with the Clinton era. This criticism is only partially supported by statistics. The level of enforcement under Lauriski was to a certain degree lower than that of the Clinton administration in general. For example, as Figure 6.4 shows, the average annual number of citations issued by MSHA between 2001 and 2004 was 115,608, 4.7% less than that in the Clinton era (1993–2000). During 1993–2000, MSHA on average issued 72,364 citations related to coal mining each year. The number was reduced 14.8% in Lauriski's tenure. The percentage of significant and substantial citations during 2001–2004 was 6% lower than that in the Clinton administration for coal mining, and 7% lower for all mining (Figure 6.6). Nevertheless, all the statistics suggest a plateau between 1996 and 2004 across the Clinton and Bush administrations (Figures 6.4, 6.5, and 6.6). In addition, although the percentage of significant and

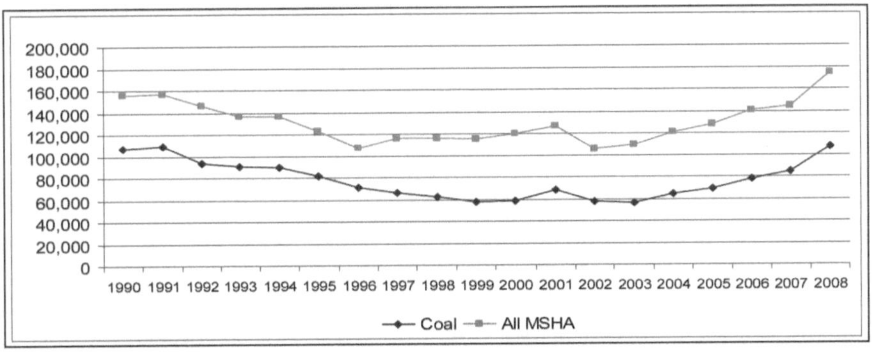

Figure 6.4 Total citations and orders, 1990–2008.
Source: MSHA (2009a).

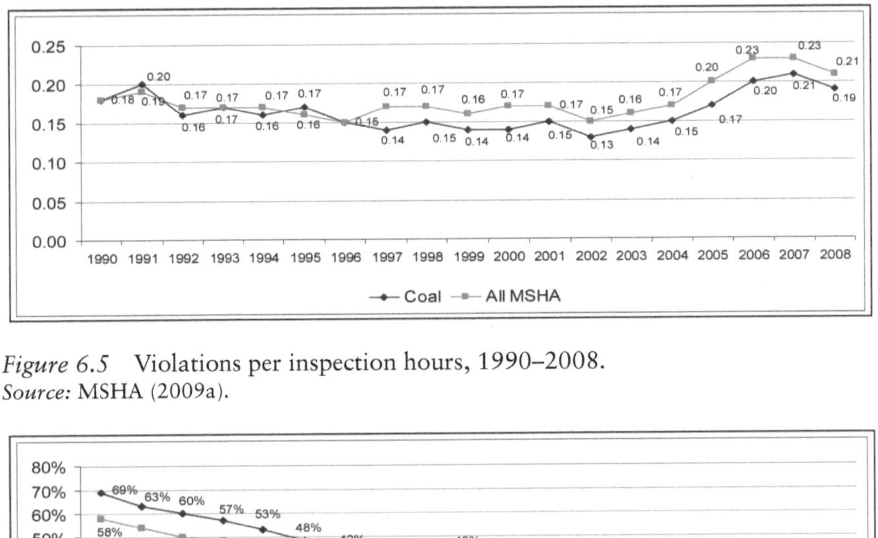

Figure 6.5 Violations per inspection hours, 1990–2008.
Source: MSHA (2009a).

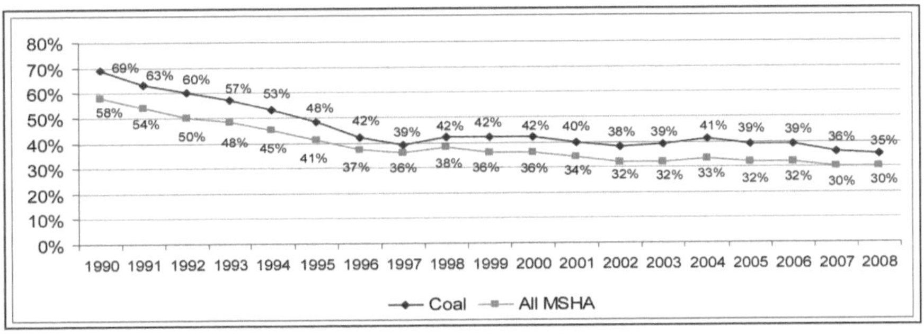

Figure 6.6 Percentage of significant and substantial citations, 1990–2008.
Source: MSHA (2009a).

substantial citations during 2001–2004 was slightly lower than that during 1996–2000, the percentage continued to drop even after 2006, when the total number of citations dramatically increased (Figures 6.4 and 6.6). Therefore, it is safe to say that MSHA under Lauriski did not significantly reduce the level of enforcement.

Some critics of Lauriski and his colleagues do have legitimate concerns because enforcement is necessary for mining safety. Even the idea of responsive regulation does not completely reject traditional command and control regulation. Rather, it emphasizes that alternative regulatory approaches cannot be effective if regulatees are not subject to the deterrence of punishment (Ayers and Braithwaite, 1992). However, the problem of many criticisms is that they see enforcement as the only regulatory tool and any move away from it as compromising mining safety. Even worse, many criticisms are designed to gain political advantage. According to an informant, some politicians in the Congress "forever" criticized the MSHA under Lauriski, and the lack of enforcement was the easiest target. An interviewee called this a "politicization of safety." During the interview, Lauriski talked emotionally about the influence of partisan politics on his efforts. He said:

> The biggest problem in this country is politics. In Washington, unfortunately, the Bush administration was very politicized, very partisan. So no matter what we did, it became politicized. That is a huge factor in how you are able to accomplish what you want to accomplish. I have very strong views about this; safety should not be politicized. UMWA, some congressional people, congressmen, and senators believe I was too close to the industry because I was an industry executive before I went to the agency. I was the first assistant secretary to have the kind of background that I had. . . . I was criticized because I have been quote "in the pocket of the industry." I was not in the pocket of the industry. That is how politics come in and play. . . . That is why you see some criticism about the things we did through my time there. But nobody really focuses on why the fatality and injury rate dropped during those four years. Nobody goes to find out why.

As mentioned earlier, coal mining safety records during Lauriski's tenure improved dramatically. However, even these statistical improvements have been questioned by some opponents. Some attributed the improvement to fewer miners working in mines and more mechanized production. Others claimed that "as a percentage of hours worked, on-the-job deaths have remained about the same over the last six years" (Ward, 2009). The purpose of this research is not to verify whether it was Lauriski's new initiatives that led to the improvement of safety performance. This task is challenging because many factors are difficult to control. Conservatively speaking, however, the MSHA's efforts moving away from prescriptive regulation at least did not jeopardize mining safety, as many have claimed.

Another respondent from the industry also noted the influence of self-interested politics. He said that Lauriski was "hammered" unfairly, and continued:

> In the U.S., the politics is always doing like this. The media is always involved. They don't care about damaging someone's name to achieve their goals. Not many people at my age would like to work in that job [the assistant secretary of the Department of Labor]. One of my friends refused to accept the position after Lauriski resigned. He said: I don't want, at the capstone of my career, my children reading a newspaper saying that I am cheating and lying.

Lauriski's disappointing experience actually played a role as an inhibitor of followers' action to move toward the risk-based regulatory model. This is because people "tend to do the things they have seen succeed and avoid those they have seen fail" based on observed vicarious consequences experienced by others (Bandura, 1986, 297).[19] Bandura (1986, 160) also argues that "foisting new practices on those who oppose them is likely to do more harm than good" because "such campaigns waste a lot of effort and resources with discouraging results that only impede later applications." A vice-president of safety at a mining company said that there would not be a partnership between MSHA and the industry any time in the near future because the MSHA leadership saw Lauriski's trouble. Coal mining companies will have to work on risk management alone. MSHA will not do it again if it cannot bring unions on board.[20]

The strong suspicions from unions and politicians in Congress described above have led Congressional committees to call MSHA officials to answer questions and to threaten cutting funding. However, another important influence of the Congress on MSHA is that the legislation it has passed significantly limits the agency's latitude in pursuing its own agenda. According to an interviewee, the language of the 1977 Mine Act itself, which primarily focuses on the physical conditions of the workplace and authorizes little discretion to the regulator, is the biggest barrier keeping them from moving toward a risk-based system regulatory model.

In addition, unlike Australia where the promulgation of legislation is controlled by "mining people and experienced people" in tripartite working groups, legislative preparation in the U.S. is largely in the hands of members of the House and Senate. As a result, promulgation of legislation in the U.S. is more open to political pressure and public anger. For example, unlike Australia, which underwent six years of serious discussion about new mining legislation after the Moura No. 2 mine explosion in 1994, the U.S. Congress passed the Miner Act just four months after the Sago disaster in response to political pressures. The West Virginia legislature passed a new law within a month of the disaster. Owing to the pressure to act immediately and the stronger influence of elected officials and lawyers compared to mining professionals, it is less likely that much new thinking and professionally sound judgments will be put into new mining legislation. Several informants from Congress, MSHA, and the industry mentioned this problem. According to an industry official:

> When you get the Miner Act that has been put on people, it was not done with sound engineering judgment. It was just done with political pressure.[21]

A congressional staffer also acknowledged that one problem in the U.S. is that legislation is often written against the backdrop of an accident like Sago. In such an atmosphere, it is very difficult to have a "reasoned dialogue at least looking at different regulatory models."[22] Such legislation, which is passed by Congress without extensive dialogue or serious considerations of different regulatory approaches, further constrains the model mongers in MSHA from pursuing regulatory changes. For instance, just because of the requirements imposed by the Miner Act, MSHA had to switch its focus back to pure enforcement after 2006.

From the Australian perspective, a government inspector was surprised at the level of intervention from Congress on legislation.

> I've been fascinated by following the stories of changing the Miner Act and how individual senators, people who don't know bugger all about mine and mine safety, people who have never been to mine sites like Kennedy, thought we were champions, they know all about it. . . . We never have lobby groups from the outside going to parliament and influencing parliament on mine safety at all. I get amazed by the sudden involvement of some Congressmen in the States, and now you cannot even appoint people to head MSHA. Stickler is still temporarily put on the side.[23]

The story described above again indicates that government regulation can not only drive, but also inhibit the growth of regulatory capitalism. Another Australian government inspector discussed why the U.S. coal mining industry is difficult to change. He thinks that it is because no one champions change in the U.S. In his view, a successful transition requires someone who initiates and champions the change and someone who is in an influential position and therefore is able to remove the barriers.[24] His observation is not completely accurate. There were people in the U.S. who champion change, such as Lauriski. The problem is that those who push for change were not influential enough to remove barriers. Their hands were tied by constraints coming from various directions.

The discussions above have shown that senior government officials in MSHA were unable to become strong model mongers as did their Australian counterparts. What has been the influence of the other two types of model mongers: mine operators and managers and union officials in the U.S.? The following two sub-sections address this question.

Mine Operators and Managers

As Chapter 5 described, the majority of mine operators and managers in Australia support the risk-based system regulatory model, although some prefer the prescriptive approach and even have challenged the duty of care provision in court. In the U.S., however, although many mine operators and managers advocate less prescriptive regulation, they do not support a risk-based system regulatory model like Australia's. This includes the lack of support from even the most aggressive companies that promote risk management. In interviews, the vice-presidents of safety at such companies expressed concerns that the Australian regulatory model is not the best because they thought that in the Australian regulatory framework the only legal requirement was doing risk assessment, and all specific requirements had been eliminated.[25] In their view, the best regulatory model should be a combination of both.[26] In other words, they still see the necessity of prescriptive standards. As an informant said:

> Regulation can tell you something probably you have not thought about. . . . Regulation is prescriptive, while risk management is performance-based. It is hard to tell if you have done a good risk management. . . . The best way to do this is set up law as the first layer and risk management [as the second layer] to work on the law requirement. Column A is the legal requirement, and Column B lists what we want to do additionally to the law requirements.

However, the biggest concern of mine operators and managers in the U.S. about the risk-based system regulatory model is the issue of legal liability. As one respondent said:

> In Australia, if anything happens at all the site, executive is responsible. Essentially, what the government says is: "you didn't consider this in your risk assessment, therefore it is inadequate and therefore you fail your duty of care requirement and therefore you are going to jail." Over here, it is "you violate the law and somebody got hurt, now you are going to jail." Duty of care is a much higher hurdle. I think that [duty of care] would cause panic in the American workforce because just that fact that you didn't recognize the hazard is criminal activity. . . . Quite honestly, we have difficulty getting people to step out in Australia and take responsibility as safety executives just for that reason alone. It can be carried too far. If it is real obvious lack of care . . . now the mine manager is at home and someone does something that they

> were trained not to do, they still hold the mine manager responsible for that. That doesn't make sense to me.

Therefore, some of the perceived benefits of prescriptive regulation and the negative consequences of the risk-based system regulatory model prevent mine operators and managers in the U.S. from becoming strong model mongers like those in Australia. Along with a mindset with "a false sense of security," the industry did not proactively promote any alternative regulatory models before 2006.[27] After the Sago disaster, the whole environment changed to become very negative to the industry. All the industry did was to defend against attacks, and, therefore, it did not have the opportunity to advocate for its suggestions. An informant from the industry explained the situation after the Sago accident:

> We were on the defensive. Honestly, our objective was trying to prevent something more damaging to the industry than where we ultimately ended up. We didn't really have the opportunity to offer suggestions that we thought would be constructive suggestions. We were trying to defend. That was the environment we were dealing with. There were a lot of members [of Congress] that even didn't want to talk to us. You are the industry that cannot be trusted; you have no credibility with us up here; you are not part of the discussion. That was the mindset of some members. Fortunately, that was not the majority of members. Even in those we could have discussion with, we didn't have opportunity to put forward suggestions.

Thus, what happened in the U.S. was different from Australia. In Australia, model mongers, particularly senior government inspectors and mine operators and managers, successfully put the risk-based system regulatory model on the agenda when the "policy window" opened because of the switch from the conservative to the labor government. By contrast, in the U.S., when the "policy window" opened after the Sago disaster, model mongers, including entrepreneurs in MSHA and mine operators and managers, were not strong enough to change the existing regulatory arrangement. MSHA's efforts led by Lauriski already had been frustrated. The industry was perceived very negatively, especially after the Sago disaster. Mine operators and managers also had mixed feelings about the risk-based system regulatory model. Consequently, the contending group of model mongers, primarily consisting of unions, took the opportunity to put their models on the agenda. As a result, coal mining safety regulation became even more prescriptive after 2006.

Unions

As the last chapter showed, unions in Australia also prefer a more prescriptive style of regulation in general. However, they are not unanimous. Some people from unions, particularly check inspectors, embrace the risk-based system regulatory model because they have the cognitive capacity to understand the value of risk management based on their experience as inspectors. These "rebels" serve as a bridge between two contending groups. They can bring unions' positions closer to those of government inspectors and mine operators. By contrast, there is no evidence of any "rebels" from unions who publicly endorse the risk-based system regulatory model in the U.S. According to respondents, some safety professionals in unions do desire a more proactive approach to safety such as accident prevention in certain areas. However, they are not in leadership positions and have no opportunity to exert influence. Moreover, unlike Australia where model mongers work in a small and closed environment, stakeholders' input for legislation and regulations is collected through public hearings in the U.S. In this more open environment, it is difficult for delegates to say anything different from the official statements of their organizations. As an interviewee described, "people from unions carry their flag wherever they go."

Certainly, some people from unions at least are not against mine operators' voluntary adoption of risk management.[28] However, they tend to become hostile if risk management is tied with any regulatory approach that might dilute the specificity of regulations. That is why the union members on the NMA's Mine Safety Technology and Training Commission chaired by Larry Grayson did not oppose its recommendation on risk management. According to Grayson, the Commission only recommended that operators do risk management and "did not say anything to put it into regulations."[29] However, as earlier chapters suggested, little room exists for operators to adopt a risk management model in a highly prescriptive regulatory environment.

The hostility of unions in the U.S. to the risk-based system regulatory model is evidently due to two factors. First, unions have vested interests in existing regulatory arrangements. According to Needleman (2000, 71), the creation of the OSHA Act of 1970 and the Coal Act of 1969 was a "hard-won victory for workers in a nation where control of work hazards had historically ranked low on the social agenda" after decades of militant social movements. This is also the reason behind the ironic result that the country with the weaker labor movement and the more influential business sector has ended up with a more adversarial regulatory system (Clark, 1999). A simple analogy is that a family that struggles to purchase a house for their entire life would do anything it can to protect its property. By contrast, a richer family may cherish its house less because it can get another one more easily.

Second, the hostility is strengthened when unions see the observed negative vicarious consequences in Australia with the shift from a prescriptive to a risk-based regulatory model. As previously discussed, some Australian companies, although have risk management in place, did not implement correctly. Observing what happened in Australia, they predict that similar outcomes will emerge if people push risk management in the U.S. According to Bandura (1986, 297), "the consequences experienced by others convey information to observers about the kinds of actions that are likely to prove effective or detrimental." The actual effects of vicarious consequences depend on observers' judgments. If people determine the contingencies in the environment to be similar, they predict that similar results would occur after similar actions. An interviewee recognized this effect of observed vicarious consequences. When explaining why unions are against risk management, he said:

> UMWA do not want to have risk management because they see what happens in Australia primarily. What happens in Australia is they have performance-based regulations. They are not specific on you have to do this, this, and this. UMWA thinks there should be more specific regulations, and therefore if that type of performance-oriented regulation is associated with risk management like Australia does, they would rather not have it. . . . They interpret that means put it into regulation that you have to do risk management. By doing that, you end up as Australia.

Hence, had the regulatory reform not happened in Australia, unions in the U.S. would have been less hostile to risk management.

As the last chapter showed, the successful shift toward the risk-based regulatory model in Australia results from not only the existence of strong model mongers, but also the strong influence of model missionaries and mercenaries, who distributed knowledge and helped different parties achieve shared understandings, through providing training and participating in accident reviews or periodic government mine safety reviews. The next sub-section will examine the role of model missionaries and mercenaries in the modeling process in the U.S.

Model Missionaries and Mercenaries

In the U.S., there are also model missionaries and mercenaries such as Larry Grayson and NIOSH researchers. However, their influence is much weaker because of the lack of channels through which to exert influence. For example, the last chapter showed that Jim Joy provides training classes on risk management to people from government, the industry, and unions. There is no model missionary or mercenary offering such a level of training in the U.S. NIOSH is the closest. It organized several risk management workshops in 2008. However, participants were limited to people from the industry. According to an informant, people from unions did not join these workshops.

> Interestingly, NIOSH had a two-day conference. They got Jim Joy. Unions never show up. They were signed up for it, but they never showed. I was told that they decided if they showed up they would be endorsing it, and they didn't want to endorse it. I go to a lot of sessions I don't endorse.[30]

NIOSH has the potential to provide a type of training similar to Joy's after it published the MHRM report. It has plans to organize workshops to disseminate risk management across the country. However, to what extent it can bring people from government and unions to these training sessions remains to be seen.

Moreover, in Australia, model missionaries and mercenaries can help different parties to create shared understandings through participating in accident investigations and periodic governmental mine safety reviews. These investigations and reviews are very important because they are part of an enactive learning process in social cognitive theory. The different interpretations of direct experiences may lead the industry in different directions.

As previously discussed, government mine safety reviews in Australia are usually conducted by a committee with representatives from government, the industry, and unions. For accident inquiries, scholars are always involved as independent members. By contrast, in the U.S., few influential periodic mine safety reviews have been commissioned by government. Moreover, in the U.S. accident investigations are often conducted separately by each party. For example, after the Sago disaster in 2006, four investigations were conducted by MSHA, UMWA, the State of West Virginia, and the International Coal Group, the parent company of the Sago mine.[31] Similarly, after the Crandall Canyon disaster in 2007, there were both the MSHA accident report and the investigative report published by the U.S. Senate Health, Education, Labor and Pensions Committee chaired by Edward Kennedy. More important, these investigations were conducted almost completely by insiders. Independent scholars were rarely involved. For example, MSHA's Sago and Crandall Canyon accident reports indicate that all investigators were MSHA employees, including district managers, supervisors, and engineers in the technical support group, and inspectors. UMWA's investigation only included people with close links to the union as investigators. The congressional investigation was conducted by congressional staff. These types of investigations may provide checks and balances. However, they do not promote channels of communications and the opportunity for independent model missionaries and mercenaries to help different parties to achieve shared understandings.

In addition, accident investigations and mine safety reviews in the U.S. rarely involve people from other disciplines. On most occasions, they are conducted purely by people related to mining. By contrast, scholars involved in Australian accident inquiries or government mine safety reviews include not only mining engineers, but also lawyers, sociologists, and specialists in general management. This may be an additional reason why the U.S. coal mining industry is much slower than its Australian counterpart to accept practices that have been adopted in other industries.

Only very recently have some government reviews started to involve independent scholars. One example is the Utah Mine Safety Commission (UMSC) appointed by the governor of Utah to review the state's role in coal mine safety, accident prevention, and accident response immediately after the Crandall Canyon disaster. According to the UMSC (2008), the commission was headed by an independent chairman, Scott Matheson Jr., who is a professor at S. J. Quinney College of Law at the University of Utah. This commission put forward some innovative recommendations, such as the establishment of a Mine Safety Technical Advisory Council including representatives of miners and operators (UMSC, 2008, 51). However, the issue of risk management was not included in the Commission's recommendations, partly because no model missionaries and mercenaries were involved.

In sum, this section shows that a major obstacle keeping the U.S. government from moving toward risk-based system regulation is that the model mongers in MSHA who championed the change were not strong enough to remove barriers. Moreover, there are not many opportunities for model missionaries and mercenaries to exercise influence. However, even if there were influential model mongers, missionaries, and mercenaries, it would be still difficult for the U.S. government to move toward a risk-based system regulatory model. This is because there are few webs of dialogue that can bring different parties together and help them achieve shared understandings.

Webs of Dialogue

The Lack of Tripartite Mechanisms

New mining statutes and regulations in Australia were to some extent a compromise among different parties. In other words, new laws held some benefits for all concerned. The compromise could be reached partly because of the existence of institutionalized and close webs of dialogue, which are mainly exercised through tripartite mechanisms. These mechanisms bring contending groups of model mongers together and encourage more genuine dialogue to search for a win-win situation.

Compared with Australia, there are few such tripartite mechanisms in the U.S.[32] Both bill preparation and administrative rulemaking rely on a more "judicialized" style of public hearings and public notice and comments. Such a style can be open to input from a wider range of social groups and individuals. However, there is often little communication among the interested parties. It is much like in a courtroom. Both plaintiffs and respondents present their arguments and evidence to an independent judge, but they never talk to each other except through a formal cross-examination process. In addition, there are not as many tripartite bodies in the U.S. as there are mine safety advisory councils in Australia, especially those that have a strong influence on legislation and regulations. Some informants noted that there are periodic tripartite meetings, but not on a formal basis in the U.S.

The only longstanding tripartite body is the West Virginia Board of Coal Mine Health and Safety, established in 1977. According to WV Code

§22A-6-1, the Board must have seven members, six appointed by the governor with the advice and consent of the state senate with three members representing mine operators and the other three representing workers. The seventh member is the Director of the Office of Miners' Health, Safety and Training. The Board has the authority to review mining rules and regulations, revise existing rules, and promulgate new rules and regulations.

The tripartite Board was put in place in West Virginia for the following reasons. First, the mining industry is very important to the state economy. As Table 4.1 shows, the mining industry contributes 5.9% of the GDP of the state, second after Wyoming in the nation. Second, mining health and safety is at the top of the political agenda because West Virginia is the largest underground coal-producing state (EIA, 2009a). Underground mining is more dangerous than surface mining. Third, West Virginia has the strongest union influence in the U.S. Right-to-work laws are not enforced in the state.[33] Fourth, the state legislature, which in the past totally controlled the health and safety agenda, found it difficult to deal with the issue because it only convenes once a year and involves a large number of people who have diverse backgrounds. As an informant noted:

> These issues used to be determined by legislature. After years and years of bitter controversial discussion, that is why they created the board, put in the hands of the experts. Because of the diverse background of the state legislature, you got doctors, lawyers, housewives, you got business people. In 1977, they said "hey why are we doing this? Why are we trying to figure out what the right thing for coal mine safety is; let's create a board."

According to interviewees, the Board has two major advantages. First, it puts mining health and safety issues into the hands of a small number of professionals, and therefore to a large degree it prevents the influence of self-interested politics.

> The purpose of the structure of the Board is to limit the amount of people representing both entities [so they can] agree on something instead of having to combat the situation in the political arena which involves 134 people. In my opinion that is why the board exists. They have the authority to get a rule, that rule can become law. You can get that done better with three people from the industry and three people from unions than 134 politicians. I think why the board was consummated, every time when we have legislative session, it is a chaotic type of situation, the industry said we want to do this, labor said we need to prohibit this. It is all political. Everybody has their agenda. The idea is the Board can function much better with a small number of people . . . it is not easy, but it is easier to do with six people rather than 134 people.[34]

Second, since the Board has equal authority with the legislature in promulgating rules and regulations, it "forces" mining operators and unions, who historically are very contentious, to sit and work together.[35] This provides an opportunity for dialogue and trust-building.

> I think those parties recognize that if anything is going to happen they have to sit down and look into each other's eyes and jointly develop it because one side of the table won't accept the other side's proposal without incorporating their own views and their own policy issues. The process is not totally devoid of politics . . . but the parties at least have the opportunity to communicate and have professional discussion without getting mad at each other. . . . The leadership on the management side of the table has worked with the same leadership on the labor side of the table for a number of years, everybody knows one another, you feel some trust and partnership.[36]

The situation in West Virginia is thus similar to that in Australia. It evidently is the importance of the industry to the economy and the existence of strong union influence that led to the formation of the tripartite mechanism. Major parties are forced to negotiate owing to the existence of a tripartite body that has concentrated rule-making authority.

Very recently, this type of tripartite body having authority to promulgate regulations has been adopted increasingly in the U.S. One example is the creation of the Utah Office of Coal Mine Safety under the Utah Coal Mine Safety Act of 2008 (Utah Code §40–2-201) following the UMSC's recommendation mentioned above. A "Mine Safety Technical Advisory Council" has been created within the Utah Office of Coal Mine Safety. According to Utah Code §40–2-203, the "Mine Safety Technical Advisory Council" consists of 13 voting members and 5 nonvoting members, including representatives from operators, unions, and government.[37]

In addition, the tripartite idea was adopted by the first major amendment to Pennsylvania's 125-year-old mine safety law, which became effective in July 2008. Following the West Virginia model, this new law creates a seven-member Board of Coal Mine Safety. The board consists of three equal representatives of mine owners and mine workers and a chairman, the secretary of the Department of Environmental Protection. The board has the authority to write new mine safety regulations.

Based on these developments, it might be predicted that this type of tripartite bodies that has authority to write regulations will continue to grow in the future. However, these new developments from a different perspective demonstrate the lack of such mechanisms in past decades not only at the federal level but also in the states.

The general lack of webs of dialogue through tripartite mechanisms in the U.S. can also be supported by other evidence. First, as mentioned, accident investigations, which are usually carried out by a committee joined by three parties in Australia, are often conducted separately by MSHA, unions, Congress, state governments, and mining companies. Second, in Australia, there are annual health and safety conferences, which are organized by a tripartite committee. Participants at these conferences come from all major parties. The conferences are considered to be a major factor in helping different parties in Australia to agree on the general direction of regulatory reform. By contrast, at least based on my observation, even safety conferences in the U.S. are split. The conferences I participated in are organized by the industry. I rarely saw people from unions at these conferences. This observation was confirmed by some informants. As one indicated, the only conferences at which he had seen both industry and union participants are some professional development conferences organized by the American Society of Safety Engineers and the National Safety Council. Even at these conferences, however, people from the industry and unions are just other attendees and they do not do anything together. The organizer of the Bluefield Coal Symposium also acknowledged that he did not think much about inviting unions to join the conference. During the interview, he said that he would give some thought to this for the next symposium because I raised the issue. However, he was still worried that some people from union-free operations would not come to the meeting if union people were asked to speak, because they do not want the union to organize in their operations.[38]

The question that arises is what factors prevent the formation of tripartite mechanisms in the U.S. coal mining industry. As Chapter 5 shows, the popularity of tripartite mechanisms in Australia is attributed to institutional foundations in industrial relations and culture, as well as political systems. These three institutional foundations also explain why relatively few tripartite mechanisms exist in the U.S.

Institutional Foundations

Industrial Relations

The form of tripartite mechanism in Australia reflects a strong labor movement and unions' direct connection with the Labor Party, which have created a political "power block" opposed to employers. By contrast, in the U.S., the political landscape is not organized exclusively around labor-management relations. This is because, first, the labor movement in the U.S. is not as strong as in Australia. Throughout U.S. history, there have been strong anti-union movements. For example, the Taft-Hartley Act was passed in 1947 as an amendment to the National Labor Relations Act of 1935 in response to the UMWA's striking during the Second World War. The Act prohibited unions' unfair labor practices and also authorized individual states to pass right-to-work laws. Moreover, there has not been compulsory voting in the U.S. as in Australia. Second, unions in the U.S. also do not have a specific political party to represent them. The Democratic Party, despite strong links with unions, is not an exclusively labor-based party. Third, as Clark (1999) shows, the labor movement in the U.S. through the 1960s and 1970s was closely connected with other social movements, such as the environmental and consumer movements. Unions formed a broad coalition with other public interest groups. By contrast, the coalition of the Australian labor movement was much less diverse. Unions were the dominant group (Clark, 1999, 101). As a result, different from their Australian counterparts, unions in the U.S. are one major group in the labor movement, but other important groups have also been involved. Added to the tradition of a strong suspicion of government authority, bill preparation and rulemaking on occupational health and safety in the U.S. are designed to provide access to a wider range of social groups, not just industry and unions.

This type of institutional design in the U.S. has advantages over the tripartite mechanisms in Australia in terms of involving more interest groups in legislative processes.[39] However, the involvement of diverse interest groups makes health and safety issues more vulnerable to the influence of self-interested politics, which prevents the two major stakeholders—the industry and the unions—from cooperating with each other. When explaining the increasingly cooperative relationship between the industry and unions in West Virginia, an informant said that cooperation is very difficult at the national level because "it is a much bigger ocean. Not all the fish can get their meat."

Adversarial Culture

In Australia, a culture of cooperation is ingrained in the coal mining industry, partly because coal fields are geographically concentrated. Moreover, the inspector group is relatively small. Each inspector is assigned to a certain number of mines for many years. Hence, it is easier for government inspectors to develop long-term and cooperative relationships with mine operators and workers.

By contrast, the culture in the U.S. coal mining industry is much more diverse, partly because coal fields are located from Appalachia to the Midwest. More important, there is much less of a cooperative culture among inspectors, operators, and unions. This is the result of both macro institutional design and the lack of cooperation at mine sites. As discussed earlier, the philosophy behind the design of the coal mining safety regulatory regime in the U.S. is to prevent "capture" by the industry. Therefore, unlike their Australian counterparts, inspectors in the U.S. are rotated among different mine sites. They also have to follow very detailed, step-by-step inspection procedures. Discretion is intentionally designed out. The primary mission of inspectors is to find fault and write citations, but not to help mine operators identify and control hazards. In this type of institutional environment, it is very difficult for inspectors and mine operators to develop cooperative relationships.

Moreover, because of the significant increase in the amount of civil penalties and the levels of enforcement after the passage of the Miner Act, the relationship between government inspectors and operators has become even more adversarial. According to an informant, "MSHA is forced to measure their success by the number of citations." Even worse, as described previously, the MSHA is recruiting more and more inexperienced inspectors to meet the needs of its workforce. As a result, the challenge rate to issued citations has increased dramatically. An interviewee noted:

> Right now the feeling about the enforcement agency and how punishing it is has risen dramatically. If you think what was written in the 1980s was a strong feeling, you ought to try to talk to them now. There's an extremely strong feeling that MSHA is being unreasonable among industry people at this point of time. They just don't understand why they've gone so far beyond where they were. That is why they are challenging all these things, all these violations that are not clear and decisive ones. Like there is no doubt I was wrong on that, so therefore we fix it and drop it. But on all those shady areas, those fuzzy ones, they're challenging every one of them. It is really really butting heads right now.

In addition, cooperation is difficult at the bottom level—in the day-to-day performance at mine sites. An informant who was a superintendent in a coal mine told a vivid story about the hostile relationship between workers and operators. The company he served wanted to examine some particular ventilation issues at another mine of the company. They invited MSHA inspectors who had specialties in that area to provide technical supports. He also joined the team with the safety director, a mine union representative, and an international union representative. Every day when they first gathered, the union representatives at the mine would tell the MSHA inspectors "come here, come here. I want you see this." They took MSHA people somewhere to show them something and said: "what are you going to do about that?" It took them one hour or two that should have been spent in investigating the issue together. This situation went on for several days. Finally, the MSHA inspectors said: "we are in technical support; we are not here to write violations on something like that. We are not going to do that anymore; we have to do this investigation."

He continued to point out that the right way to do business should go beyond each person's concern and deal with issues for one another. He said:

> If you just focus on the little things that *your mission as a manager and your mission as a union health and safety representative, or your mission as MSHA inspector and it is driven from the top, you never have a chance to come together.* All things get solved in a good way by

> those who cooperate at the basic level, at the mine level. If that filters all the way up through the organization, the whole company is that way, then it happens at the company level. If all the companies try to work with other labor organizations and effectively put together a good relationship in West Virginia, Pennsylvania, Virginia, etc., then it becomes easy to happen at the national level. We are broken at the base level right now [emphasis added].

In the U.S. coal mining industry, both the institutional design at the macro level and day-to-day performance at the micro level have created an environment in which people tend to care only about their own concerns and forget others' concerns and the common good of the whole industry. That is why cooperation in the U.S. coal mining industry is so difficult.

Fragmented Political System

The establishment of the webs of dialogue in Australia is forced to some extent because the operators and unions do not have allies in the parliament, which has little influence on bill preparation and rulemaking. Hence, the major parties have to sit together to negotiate. The scenario in the U.S., particularly at the federal level, is totally different. The Congress controls legislative activities and has a very strong influence over MSHA's rulemaking. Both the industry and unions have strong connections with members of Congress. Therefore, both sides tend to use their political leverage in the Congress when they find it difficult to reach agreement, instead of sitting at the table to continue negotiations. According to informants, this kind of activity happens very often when discussing issues such as coal dust control, the movement of equipment, and the use of diesel. When talking about the increasingly cooperative working relationship in West Virginia, an interviewee said:

> Washington, DC, is a different place. It is very polarized inside the Beltway. You have national party politics, which very much control the tempo and the issues. It is much harder [to cooperate].

Therefore, the fragmentation of government institutions in the U.S. opens many avenues for the industry and unions to pursue their interests. As a result, it is difficult for them to sit down and work together. West Virginia is probably an exception in the U.S. According to informants, there has been a lot of improvement of in the industry-union relationship in West Virginia. They have seen "very professional, honest, and open" discussions, dealing with puzzles they have rather than politics.[40] The informants attributed this improvement partly to the tripartite Board. Since the state legislature has authorized significant legislative power for the Board, it is often the first place that the two parties approach. As a result, there are much fewer "bypassing" activities, although they do exist because the state legislature still has the authority to pass laws. As an informant indicated, often times, the bypassing activities are reversed. In other words, people go to the Board when they cannot work out anything in the legislature. In addition, given the existence of a powerful board in terms of writing and changing regulations, people are reluctant to conduct those kinds of "bypassing" activities because they are worried about undermining the trust between the two parties.[41] This is because both parties have to gain the support of the other in order to have anything passed by the Board.

CONCLUSION

This chapter has discussed the three main barriers that prevent the U.S. government from moving toward a risk-based system regulatory model: the lack of (1) strong imitative pressures from general OHS regulations, (2) influential model mongers, missionaries, and mercenaries, and (3) webs of dialogue that can bring different parties together. However, a central problem behind the second and third barriers is that the institutional design in the U.S. is based on distrust and checks and balances. This has created fragmented government institutions that can open many avenues for a wide range of interest groups to participate in policy-making. As a result, many issues become unduly politicized. Each party focuses only on its own interests and pays little attention to others' concerns. Instead of pursuing a common goal, each party tends to advance its interests by weakening others. This kind of mentality has imposed enormous pressures on the model mongers in MSHA who moved away from prescriptive regulations; precluded opportunities for model missionaries and mercenaries to bring fresh ideas and knowledge into the industry; and undermined the basis on which webs of dialogue can be formed. In the U.S., there is a lack of a central and powerful entity like the tripartite legislation review group in Australia and the Board of Coal Mine Health and Safety in West Virginia, which can force the major stakeholders to sit together and have more professional, genuine, and respectful dialogue. Compared with Australia, the U.S. finds it more difficult to create what Braithwaite (1993, 87) called for—a policy space where interest groups mutually respect each other. He said:

> We have a greater chance of efficient and effective regulation if we have a regulatory culture where regulation reviewers and consumerists actually listen to each other and respect the concerns of the other; we have a lesser chance of cost-effective regulation if these two constituencies see their mission as to the destroy the other, taking it in turns to win battles without either side winning the war.

Finally, the research in this chapter confirms the findings stated at the end of last chapter. The Braithwaite and Drahos (2000) framework about the reactive sequence of regulatory change needs to be modified. As described earlier, individual entrepreneurs within the state who advocate regulatory innovations, such as Lauriski in MSHA, can initiate regulatory changes proactively without the occurrence of major disasters. Therefore, a modified analytical framework of regulatory change at the governmental level appears in Figure 6.7.

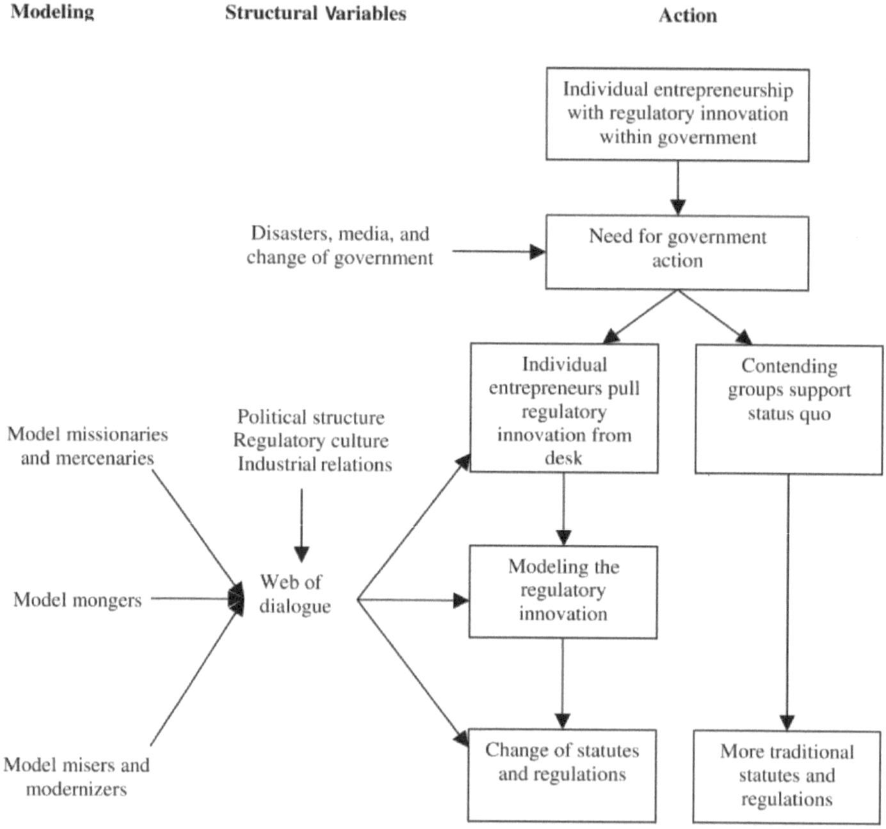

Figure 6.7 Process of driving up standards and modeling at the government policy level (modified).

7 Conclusion

INTRODUCTION

As Chapters 1 and 2 described, the major theoretical purposes of this book were twofold. The first was to fill the gap in current literature about the development of risk-based governance by examining concrete mechanisms that drive, and, more important, block its diffusion in a particular setting—coal mining safety regulation. Since risk-based governance is a key element of regulatory capitalism, this study also helps one better understand what drives and constrains the growth of regulatory capitalism. The second purpose was to apply the Braithwaite and Drahos framework, particularly the theory of modeling. This chapter will first summarize the findings regarding both the drivers of and barriers to the diffusion of risk-based governance, and discuss the contributions of this research to the Braithwaite and Drahos framework. Practical implications will then be addressed. The chapter ends by discussing implications for democratic governance in an era of regulatory capitalism.

DRIVERS OF AND BARRIERS TO THE DIFFUSION OF RISK-BASED GOVERNANCE

This research has identified several factors whose existence or absence served as either a driver of or a barrier to the diffusion of risk-based governance. These factors also have implications for studying the rise of regulatory capitalism more generally.

Common Fate

As previously discussed, modeling is not an isolated process. Rather, it is the next logical step of a change of mentality and part of the industry's collective process of driving up standards. Modeling best practices may occur in some individual companies that have built up a higher level commitment to safety. These companies are, however, unlikely to take collective actions until a common fate is recognized.

The introduction of risk management in the U.S. coal mining industry resulted from the emerging recognition of a common fate among some of the largest coal mining companies after the Sago disaster. This recognition was a response to the threats of increasing government regulation and public actions against the industry. Because of this recognition, the industry, led by the largest companies, created or enrolled organizations (e.g., industry associations, international bodies, and research institutes) to search for new models to improve their reputational assets. Nevertheless, the feeling of a common fate in the U.S. is much weaker compared with that in Australia, because the U.S. coal mining industry is less internationally oriented and less important to the national economy. Consequently, the American coal mining industry is somewhat less concerned about its reputation and therefore less motivated to adopt world best practices, such as risk management.

Thus, the extent to which a feeling of common fate exists in an industry is a factor affecting the diffusion rate of the new order of regulatory capitalism in general and risk-based governance in particular. In addition, the process of modeling risk management in the American coal mining industry illustrates the reciprocal causation among different components of regulatory capitalism. It is the increase of pressures imposed by government regulations, civil society, and the public that leads to the growth of industry's self-regulation.

Prescriptive Regulation

The style of government regulation may provide incentives or disincentives for regulated companies to establish internal risk management systems. In this case, highly prescriptive regulation in the U.S., which increased after 2006, was an inhibitor to the diffusion of risk management. On the one hand, compliance has consumed the resources that companies could have devoted to adopting the risk management model; on the other hand, such regulation has created a compliance mentality in the industry that is not compatible with the idea of risk management. Hence, government regulation is a double-edged sword in regulatory capitalism. The increase of government regulation can encourage or discourage the establishment of a self-regulatory regime in an industry.

Homogeneous Industrial Structure

Consistent with the findings of some existing research, the diversity of the industry is a major barrier to the American coal mining industry's adoption of the risk management model. On the one hand, the diversity leads to different levels of commitment. On the other hand, many small operators lack the resources to develop highly complex risk management systems. As such, it is very difficult to bring everyone together to drive up safety standards.

This research suggests that risk-based governance is more easily adopted in an industry where large companies are dominant. However, even large companies that have built up a stronger commitment to safety may adopt different models of self-regulation.

Competing Models

This research finds that one barrier to the diffusion of the risk management model in the U.S. is the existence of a competing model—BBS. Many American coal mining companies find it easier to accept BBS because (1) union density is low; (2) people's cognitive patterns tend to focus on risk at the behavioral level; and (3) BBS is more popular, less complex, more compatible with American individual and litigious culture, and more likely to produce immediate and observable results. The implication of this finding for regulatory capitalism is that the existence of competing self-regulatory models can also prevent the diffusion of regulatory capitalism, particularly when a new model that embodies the new order of regulatory capitalism (risk management in this case) encounters an more traditional model, such as BBS.

Imitative Pressures

This research also suggests that imitative pressure, particularly the pressure from a closely linked policy field, is an important factor that determines whether a regulatory model can be adopted in a sector. As Chapters 5 and 6 showed, the development of coal mining safety regulation in both Australia and the U.S. basically followed the path of general OHS regulations. The difference is that Australia followed risk-based system regulation as opposed to an intermediate form between risk-based system regulation and command and control regulation that MSHA followed in the U.S. This also illustrates the important role of diffusion in the growth of regulatory capitalism.

Strategic Actors

Diffusion involves not only imitation, but also cognitive learning and interpretation, as well as political struggles. Thus, there must be strategic actors to initiate and champion change and resolve the problems encountered in the modeling process. Whether such actors exist to a large extent determines whether modeling can be successful. These strategic actors include model mongers, missionaries, and mercenaries.

As Chapter 3 showed, it is model missionaries and mercenaries who introduced the model of risk management and spread knowledge about the model in the U.S. on the organizational platforms enrolled or created by some of the largest companies that have built up greater commitment to safety. On the demand side, it is the different interpersonal connections and cognitive patterns of model mongers (e.g., vice-presidents of safety) that largely determined the different models adopted by the two American companies. At the government policy level, risk-based system regulation has been successfully adopted in Australian coal mining safety regulation because of the strong influence of model mongers, missionaries, and mercenaries. By contrast, in the U.S., model mongers who pushed for change were constrained by forces from various directions. Model missionaries and mercenaries lack channels to exert influence. This research suggests that in studying the diffusion of regulatory capitalism in the contemporary era, focus has to be concentrated on knowledge actors (e.g., model missionaries and mercenaries) and the organizations behind them.

Webs of Dialogue

The findings of this book also indicate that the study of modeling cannot ignore the influence of structures, particularly webs of dialogue. Since modeling involves cognitive learning and interpretation as well as political struggles among contending groups of model mongers, missionaries, and mercenaries, successful modeling requires webs of dialogue that can bring different parties together to built trust, seriously discuss and interpret new models, and reach shared understandings. As Chapters 5 and 6 discussed, the lack of webs of dialogue in the U.S. has made the modeling process more difficult. This finding suggests the existence of webs of dialogue can be an important factor that influences the diffusion of a new regulatory idea and practice, such as risk-based governance, in regulatory capitalism.

Among the factors listed above, some already have been supported by findings in other research, such as the influence of common fate, industrial structure, and government regulation. Others are still questionable as to what extent they can be generalized to other contexts (e.g., the existence of competing models and webs of dialogue). Despite the limitation on generalizability, these factors at least can draw the attention of researchers who study similar topics in the future.

CONTRIBUTIONS TO THE BRAITHWAITE AND DRAHOS FRAMEWORK

Although the analysis followed Braithwaite and Drahos's concepts and framework in studying regulatory globalization, this research did not follow them rigidly. In other words, I accepted the bottle of wine, but also added something new. As Braithwaite and Drahos (2000, 581) note, their theory of modeling has "micro foundations from both psychological work in the social cognitive tradition (Bandura, 1986) and sociological literature in the symbolic interaction tradition (Burke, 1945; Duncan, 1969; Mead, 1950; Turner, 1974)." However, their purpose is not to "make a contribution to the understanding of these foundations, but to build on them" (*ibid.*). Braithwaite and Drahos added a political element—the politics of empowerment—to the foundations. Without dismissing the value of the politics of empowerment, my research goes back to one of the foundations and gives greater emphasis to the social cognitive element. This increases the explanatory power of the theory of modeling for what drives and resists regulatory changes in two primary ways.

First, the theory of modeling adopted in this research directs attention to the impact of individuals' cognition on regulatory changes. This is particularly important when studying diffusion in the era of information capitalism. As the research finds, senior government inspectors became strong model mongers of risk-based system regulation in Australia because of their cognitive capacities based on their experiences as mine managers. By contrast, the general lack of such cognitive capacities not only prevents American government inspectors from being model mongers, but also makes them inhibitors of change. Likewise, unions in Australia agreed to the general direction of regulatory change because their check inspectors had the cognitive capacity to understand the value of a risk-based system regulatory model. Since people's cognition is crucial to the change of regulatory orders, particular attention has to be given to individuals who shape other people's cognitions. Model missionaries and mercenaries have important influence. They can create and distribute knowledge related to a model to a broader audience. When knowledge that has been popularized by model missionaries and mercenaries convinces a certain number of people, particularly those who have control over the state's capacity, the model is likely to gain a hegemonic position. As Chapters 5 and 6 showed, the different levels of influence of model missionaries and mercenaries explain why risk-based system regulation has been successfully modeled through in Australia, but not in the U.S.

Second, the theory of modeling in this research identifies the influence of competing models. Because of an emphasis on the politics of empowerment, Braithwaite and Drahos (2000) see modeling as a process of political struggles between contending interest groups that advocate oppositional models. This research, however, suggests that people may adopt different competing models, even though they come from the same interest group (e.g., larger companies that have built up higher levels of commitment to safety in this case). Politics is certainly a factor that determines which model to adopt. For instance, BBS is popular in the U.S. partly because of the weaker influence of unions. However, other social and psychological factors also influence which model is more likely to be introduced, such as interpersonal connections, people's cognitive structure and capacity, and the properties of the model itself.

Moreover, the research suggests that the Braithwaite and Drahos framework for driving up standards needs some modifications to better explain actual regulatory changes. First, the proactive and reactive sequences of regulatory change in the framework may each include features of the other. For example, the research finds that the so-called proactive and bottom-up process of driving up standards, which starts from entrepreneurs in the industry or companies, is also reactive to disasters. At the same time, there are also proactive factors in the so-called reactive and top-down sequence of regulatory change led by the state. Governments not only react to disasters; they may also act proactively when control of government changes, for example, when the Labor government in Australia and Lauriski in the U.S. came into office. Second, the research finds that model mongers may not always have multiple models to propose, as Braithwaite and Drahos (2000) suggest. They may have only one model when imitative pressure is strong.

The Braithwaite and Drahos framework, particularly the theory of modeling, is certainly not the only theory that stresses the influences of people's cognition. Some often-used theories of policy process, such as Kingdon's (1995) multiple streams and Sabatier and Jenkins-Smith's (1988) advocacy coalition framework (ACF), are also sensitive to the cognitive aspect of policy change. Although it does not say explicitly, Kingdon's theory of multiple streams addresses the significance of cognition in at least two aspects. First, it emphasizes the selective role of policymakers' cognition in the ideas generated by individuals in policy communities, which refer to networks of specialists who share a common concern in a policy area. Not every idea floating in policy communities receives the equal attention of policymakers. Only those that are more compatible to policymakers' value and cognition are developed into policy proposals. Second, cognitive factors have impact on stream coupling when policy windows open. To couple streams, policy entrepreneurs often need to manipulate the dimensions of policy issues to match solutions to problems and to persuade their audience of policymakers. The ability of policy entrepreneurs to manipulate dimensions is constrained by their existing cognitive patterns.

Compared with multiple streams, ACF pays more attention to the impact of cognition because it clearly differentiates cognitive from non-cognitive aspects of policy change. Non-cognitive factors refer to events that occur outside policy subsystems, such as changes of socioeconomic conditions and governing coalitions. The cognitive aspect of policy change means a change of belief systems through policy-oriented learning both within and between coalitions. Since learning is often filtered through preexisting beliefs, cross-coalition learning is very difficult. Therefore, ACF emphasizes the role of professional forums in facilitating cross-coalition learning.

It can be seen that both multiple streams and ACF share some similarities with the theory of modeling. For example, the activities of stream coupling conducted by policy entrepreneurs in multiple streams reflect the model mongering process in the theory of modeling. Professional forums in ACF play the same role as model missionaries and mercenaries do in the theory of modeling: to help different contending groups or coalitions to achieve shared understandings.

However, while acknowledging that multiple streams and ACF are more comprehensive theoretical frameworks, the failure to use modeling as a central organizing concept in these theories gives the theory of modeling a superior position in providing a more accurate explanation on policy change. This is particularly the case in the contemporary information society in which the creation and diffusion of information have become critical parts of activities. As a result of the rampant information creation and diffusion, on most occasions, "ideas" generated in policy communities in multiple streams and "beliefs" held by coalitions in ACF are developed based on the interpretation of existing models. Moreover, using modeling as an organizing concept also provides researchers a clear source where they can explore abstract "beliefs" of actors and groups in policy inquiries. The source is those actors and groups' cognitive interpretations of existing models. How the theory of modeling differs from and how it can be integrated with other alternative policy theories, therefore, merits further studies.

PRACTICAL IMPLICATIONS

In addition to the theoretical contributions, this research has practical implications for the coal mining industry and relevant policy-makers. During the time of this study, the global mining industry has been undergoing a transition led by some of the largest multinationals. As Chapter 3 described, these companies are moving aggressively toward sustainable development in response to several fundamental threats to the industry. One of their strategies to achieve sustainable development is to uphold the risk management principle. Although we have seen a similar process in the U.S., it is limited to several of the largest mining companies. U.S. companies are also largely excluded from the discourse and standard development related to sustainable development and risk management at the international level. As a result, the U.S. mining industry has ceded the leadership position in health and safety and other sustainable development issues to other countries, such as Australia and the U.K. As an informant said, "we have always been leaders, and now we start to take the back seat."

The findings of the research suggest some possible strategies for people in the industry, governments, and academy who wish to help the American mining industry to gain leadership in the world by moving toward risk-based governance. The following strategies can be considered. The first is to cultivate more strategic actors, such as CEOs and model mongers in government and companies, who have foresight and a strong commitment to safety. These strategic actors, as sparks that start fires, can initiate and lead changes.

Second, when strategic actors initiate changes, it is better to create tripartite or even multipartite mechanisms to increase dialogue, trust, and shared understandings among all stakeholders. The West Virginia Board of Coal

Mine Health and Safety and the Mine Safety Technology and Training Commission mentioned in earlier chapters are good examples. These tripartite or even multipartite mechanisms form what some scholars have called nodes of governance (Braithwaite, 2008; Burris, Drahos, and Shearing, 2005; Wood and Shearing, 2007).[1] If risk-based governance emerged during the deliberations at more and more nodes at the local level, this approach can "bubble up" to the national policy level (Braithwaite, 2008, 107).

A third strategy is to incubate more model missionaries and mercenaries to increase the depth of expertise and provide more channels for them to influence people's cognition, such as inviting them to join accident investigations or government regulatory reviews. These model missionaries and mercenaries can bring new ideas and practices to the industry.

Finally, particular attention needs to be given to the capacity of government inspectors, especially their cognitive capacities. As Braithwaite (2008, 205) states, in an era of regulatory capitalism, we need "a society with a strong state, strong markets, and strong civil society, where the strengths of each institution enable the governance capabilities of other institutions." As this research shows, the case in the U.S. is the opposite. Because of the lack of cognitive capacities, government inspectors themselves are a barrier to change. They do not enable and even undermine businesses' capacity to regulate themselves.

IMPLICATIONS FOR DEMOCRATIC GOVERNANCE IN THE ERA OF REGULATORY CAPITALISM

The last chapter shows that the highly prescriptive and adversarial regulatory regime in the U.S. has prevented risk-based governance, an embodiment of regulatory capitalism, from diffusing in the U.S. coal mining industry. The new order of regulatory capitalism, however, is not what one can choose, but a condition that one must work with. This is exemplified by the discussion in Chapter 3. Despite the unsupportive regulatory environment, the involvement of non-state actors in regulatory processes still increases. The largest mining companies have proactively adopted or are in the process of introducing internal health and safety management systems including risk management and BBS. They also work collectively to model risk management at the industry level.

What is the implication of the new order of regulatory capitalism for democratic governance? Because of the increasing involvement of non-state actors, governance in the era of regulatory capitalism has become less hierarchical and statist, and more horizontal and networked. In such a more networked order, a prescriptive and adversarial regulatory regime, which is designed largely based on the principles of representative democracy, can hardly provide a "rich democracy" in which people can be fully protected against arbitrary exercise of power (Braithwaite, 2008).[2]

According to Braithwaite (2008), Pettit's neo-republicanism fits better with networked governance than does electoral democracy. The central idea of neo-republican theory is freedom as non-domination (Pettit, 1997). To achieve non-domination, neo-republicans advocate that the electoral side of democracy is not enough to protect citizens from the arbitrary exercise of power. Citizens must be provided opportunities to contest others' domination at both national and local levels. Therefore, Braithwaite (2008, 204) claims that, to obtain a "rich democracy" in the era of regulatory capitalism, representative democracy should be complemented by contestatory and deliberative democracy at strategic nodes of governance.

According to neo-republican principles, the prescriptive and adversarial regulatory regime in U.S. coal mining safety regulation has considerable drawbacks in pursuing contestatory and deliberative democracy. For contestation and high-quality deliberation to be possible, a regulatory regime must provide not only institutional structures that bring interested people together but also adequate knowledge and information to make deliberation meaningful.

This research finds that the prescriptive regulatory regime prevents the diffusion of risk-based governance in the U.S. Since diffusion is a process of knowledge and information transfer, the regulatory regime actually serves as an inhibitor of knowledge transfer and creation. According to neo-republican theory, knowledge is very important to democracy because it can empower individual citizens or social groups, so that they can not only self-govern better, but also free themselves from the domination of others. For contestation to be possible, citizens must not only have sufficient financial resources and social and legal status, but must also be armed with adequate knowledge and information. Without such knowledge and information, when individuals or groups enter deliberation, they often face a dominant discourse that has been constituted by their opposition at their expense. To achieve meaningful contestation and authentic deliberation, therefore, citizens need to be provided alternative ideas and practices outside the current institutional structure, so that they can challenge the status quo.

Recognizing the importance of knowledge and information in the contemporary era, examining the epistemic power of various democratic institutions has increasingly drawn scholars' attention (Ober, 2008; Anderson ,2006). For example, in *Democracy and Knowledge*, Ober (2008) argues that Athens outperformed its city-state rivals at least in part because of the superior capacity of its deliberative and participatory institutions in effectively aggregating social and technical knowledge that was distributed across its large and diverse population for problem-solving.

Using epistemic democracy as a reference, the adversarial and prescriptive regulatory regime in the U.S. certainly has less power in generating and spreading common knowledge than do more collaborative regimes, such as the one in Australia. For instance, highly prescriptive regulations discourage mining companies from learning about and developing risk management systems. Government inspectors, who are hired to "go by the book" and find faults, do not have the cognitive capacity to grasp and spread knowledge about risk management. In addition, owing to the distrust between major stakeholders, few webs of dialogue allow model missionaries and mercenaries to distribute knowledge to all the parties. As a result, the industry and government regulators are not fully aware of alternative solutions to contest the current arrangement, such as risk-based governance.

Here arises anther topic for future study, examining what kinds of institutional design have more epistemic power and, therefore, promote high-quality contestatory and deliberative democracy in a large-scale country, such as the U.S. Based on the findings of this research, some tripartite mechanisms, including the tripartite legislative review group in Australia and the West Virginia Board of Coal Mine Health and Safety in the U.S., may be good candidates for case studies. Such studies would also address what role "regulocrats" should play in achieving more democratic governance in the era of regulatory capitalism (Levi-Faur, 2008, vii). Borrowing insights from the literature that legitimizes public administration in democratic governance following discourse theory, critical theory, and a pragmatist tradition (McSwite, 1997; Box, 2001), regulocrats should serve as educators and facilitators of deliberation. They should be able to bring different interested parties together, empower them by distributing knowledge or facilitating knowledge transfer, and free participants in the deliberation from being dominated by historical discourses.

Appendix A

METHODOLOGY

The coal mining industry was a completely new area for me before I conducted this research. It took me two and a half years to finish this research. The most difficult part is to get access for field studies, not only because I was an alien to the industry at that time, but also because the culture of the industry itself is very closed. Hence, I think that it is necessary to share my experience with readers. I would be very delighted if any lesson can be drawn from my experience by scholars who undertake similar exploratory studies in the future. My discussion includes three sections: research design, data collection, and limitations.

Research Design

The type of research method a study employs to a large extent depends on the purpose of inquiry. The purpose of this research is to explain a seemingly puzzling and little-known phenomenon: why risk-based governance diffuses more slowly in American coal mining safety regulation than in many other countries. Following the micro-macro analytical framework described earlier, this research seeks to explore complex dynamics between individuals and structures. Therefore, it focuses not only on outcomes, but also on processes, such as how standards are driven up in enterprises and the industry led by individual entrepreneurs; how risk management is modeled across countries, industries, and individual firms; and how modeling is facilitated or restrained by structures. Hence, for a study of this sort, qualitative methods are the best choice. As Marshall and Rossman (1999, 53) claim, the strengths of qualitative studies are demonstrated in research that is exploratory, delves in-depth into complexities and processes, addresses little-known phenomena, and involves informal and unstructured processes in organizations.

Moreover, since only several of the largest coal mining companies have adopted risk management in the U.S., a quantitative analysis based on a large sample is out of the question. So is rigorous experimentation that requires deliberate manipulation of contextual conditions because this research aims to explore causal relationships in the real world where I have no control. In this situation, a case study method is suitable. As Yin (2009, 18) indicates, a case study is preferred when investigators want to understand real-life phenomena entangled with contextual conditions, which cannot be controlled by investigators and deliberatively separated from the phenomena.

Case study research includes single- and multiple-case studies (Yin, 2009, 19). This research adopted a mixture of single- and multiple-case designs according to the different sub-questions to be answered. For sub-question 1 (how and why has the risk management model been introduced in the coal mining industry in the U.S.?), a single-case study was employed (examining the American coal mining industry and an exemplary company). For sub-questions 2 and 3, a comparative case studies design—a distinctive form of multiple-case studies—was used. This is because these two sub-questions seek to examine factors that have constrained the diffusion of risk-based governance in the American coal mining safety regulation. Had a single-case study method been adopted without a comparison with cases where risk-based governance has been more successfully introduced, there would have been no way to tell if the dependent variable—extent of diffusion of risk-based governance in the U.S.—was influenced by the independent variables under consideration or other independent variables. Therefore, comparative case studies can enhance the internal validity of the study compared with a single-case research.[1]

According to Lijphart (1971), at the heart of a scientific investigation is the intention to establish a casual relationship among two or more variables while controlling all other variables that may affect the investigated relationship. In this sense, a comparative case study is no different from experimental and statistical methods. The difference is how to deal with the problem of control. In the experimental method, control is undertaken by placing subjects into experimental and control groups. Statistical analysis achieves control by means of partial correlation, regression, or other techniques. By contrast, the comparative case study method establishes controlled relationships through case selection.

Case selection can follow two different strategies: a most similar systems design versus a most different systems design (Przeworski and Teune, 1970). The former design is also referred to as a "comparable (similar) case strategy" by Lijphart (1971). It means that the cases selected should be as similar as possible to minimize the number of relevant explanatory variables and should differ on the value of the dependent variable and the explanatory variables under consideration (Kaarbo and Beasley, 1999). In contrast, in the most different systems design, researchers choose cases that are as different as possible and try to identify similarities among cases.

This study chose the most similar systems design. Comparisons here were made at three levels: enterprises, the industry, and government policy. At both the industry and government policy levels, Australia was compared to the U.S. Both countries are developed economies with federal political systems, and both share an Anglo-American tradition. They are also among the largest coal producers in the world, ranking second and fourth in terms of coal production (Figure 4.1).

For comparison at the enterprise level, the research compared two American companies. During the time of this research, I expected that only a handful of coal mining companies in the U.S. had adopted or were in the process of developing risk management systems. Therefore, I planned to select one such company to compare with a company that had not adopted a risk management system. This research did not compare American and Australian companies for two reasons. First, it was difficult to find similar companies in the two countries. Most Australian coal mining companies are large multinationals, whereas most American companies are smaller domestic ones. Although there is one American-based multinational, Peabody, a comparison of Peabody with its Australian counterparts would have limited generalizability because the majority of American companies are not multinationals. Second, it was not feasible to conduct field studies in Australia.

I identified the candidates representing companies that had adopted or were in the process of developing risk management systems by asking informants in preliminary interviews which had the best risk management systems among American companies. After four preliminary interviews, five candidates emerged.[2] These companies can be put into two different categories. Two of them are multinationals. They are either Australia-based multinationals operating in the U.S. or U.S.-based multinationals that have operations in Australia. These two companies perhaps have the best risk management systems in the U.S. because of their direct connections with Australia. These two companies, though, were not appropriate cases for this study because the majority of U.S. mining companies fit in the second category—domestic companies. Hence, the other three were better candidates. They are among the largest coal producers in the U.S.

I started to approach the three candidate companies in May 2008 by contacting people suggested by my preliminary interviewees. One company immediately rejected my request, saying that it did not have time to accommodate my research. After making dozens of phone calls to the other two, most of which turned into disappointing voicemails, I visited their offices in August 2008. After talking with vice-presidents of safety in both companies, Whitesburg Coal finally agreed to participate. The company's name is fictitious to preserve its anonymity.

The selection of the company that had not adopted risk management systems, although guided by theory, happened somewhat by luck. Until August 2008, I had no clear idea how to approach those companies. Given my painful experience of approaching those companies having the best risk management systems, I expected that it would be even more difficult to convince a company that had not adopted risk management system to participate. Nevertheless, I knew that an ideal candidate should be among the largest coal producers to be comparable to Whitesburg Coal. This is because the literature suggests smaller companies find it more difficult to adopt risk management systems than larger ones because of limited expertise and resources.

At the end of August 2008, I participated in an industry event, in which the Vice-President of Safety of Blacksburg Coal (fictitious name) gave a presentation on its behavior-based safety program. Its program was very interesting to me, so I talked with him after the presentation and asked if he would participate in my research. To my surprise, he immediately agreed and promised me good access. Although Blacksburg Coal is not a candidate that I carefully selected, it meets the requirement of being similar to Whitesburg Coal. Both are among the largest coal producers in the U.S. They are engaged mostly in underground mining in the Appalachian area. Whitesburg Coal has introduced risk management to some extent, whereas Blacksburg Coal has not.

For qualitative research like this, there are three major methods of collecting research data: asking questions, observing events, and reading documents (Bassey, 1999: 81; Maldonado-Maldonado, 2004:30). This study used all three methods. I mainly gathered observations when I joined two industry events, two companies' internal safety meetings and a meeting of the West Virginia Board of Coal Mine Safety and Health. Interviews were conducted with people from the industry, government regulators, legislatures, unions, and the academy in both countries.

Before explaining the detailed strategies and process of data collection, an important question needs to be discussed—accessibility—because the question of access existed throughout the stages of case selection and data collection. Based on my experience in this research, people from universities, government regulators, and industrial associations were relatively easy to get access to. Gaining access to people from NIOSH in the U.S. was more difficult. Coal mining companies and labor unions were the most difficult to gain access to.

I used two major strategies to deal with the question of access. First, I started to contact academics. They are the easiest to access because they are relatively open, have fewer economic interests involved, and are sympathetic to others' research. My access to Whitesburg Coal was aided by a reference given by Larry Grayson, a professor at Pennsylvania State University. Second, I used observation not only as a method of data collection, but also as a strategy of getting access. This strategy is very cost-effective because researchers have the opportunity to meet at an event many candidate interviewees whom investigators would otherwise take much more time to contact one by one. In addition, when meeting interviewees in a place like a coffee shop, it is more difficult for them to reject a researcher's invitation to talk. For example, I failed to make an appointment to talk to a director in NIOSH many times, but I met him at Minexpo in September

2008. At first, he still declined to participate in an interview after his presentation, saying that his schedule was fully booked. However, the next day, I met him when he was hanging around the booth of an exhibitor holding a cup of coffee in his hand. I asked him again and, finally, he talked for 15 minutes with me.

The assurance of confidentiality and anonymity to interviewees is a strategy of access often mentioned by researchers. My experience is that this strategy seems not as important as many investigators suggest. No informants in my research refused to participate at the beginning and changed their minds after I told them the report would be anonymous and collected information would be kept secret.

Despite my effort, access to the two U.S. companies that I studied was not deep. In particular, only two informants from Whitesburg Coal participated in my research. It was extremely difficult to arrange interviews. This is one of the limitations of this research. The validity of the information in the Whitesburg Coal case is to some extent questionable. However, this limitation is acceptable given that the purpose of investigation is not to find out how the company implements risk management in its daily operations, but to explain how and why it has moved in that direction.

Data Collection

I employed three data collection methods in this research: observation, interview, and document analysis.

Observation

Compared with other qualitative data collection methods, the advantage of observation is that it allows researchers to discover people's patterns of behavior and relationships in a natural setting with minimum interference from other observers (Marshall and Rossman, 1999). The events I participated in included two industry events (the Bluefield Coal Symposium in August 2008 and the Minexpo in September 2008), two safety meetings at Blacksburg Coal in October and November 2008, and a meeting of the West Virginia Board of Coal Mine Safety and Health in March 2009. The two industrial events lasted around three days each. The two safety meetings and the meeting of the Board were each about a half-day in length. These short visits certainly were not intensive. Hence, observation is an ancillary approach in this research but not a major source of data.

In addition to collecting data, using this technique had three other purposes. First was to get access to interviewees, as previously stated. Second was to identify issues for study. This is necessary for an exploratory study like this because, at the early stages of the research, I did not have a clear picture about the development of risk-based governance in the coal mining industry. As such, this approach was used mainly at the early stages of this research. For example, an important finding I obtained through observations at the two industrial events was recognition that many American companies are interested in behavior-based safety (BBS), and many people in the industry do not understand clearly the difference between BBS and risk management. This observation turned out to be a major issue that I examined—a competing model of risk management. The third purpose was to cross-check interview data, through both observation and informal short conversations with people I met at the events.

Interviews

Interviews are the most important source of data in this research. This method is appropriate for two reasons. First, interviews provide an effective way to gather a large amount of data in different locations in a relatively short time. Second, and more important, interviews involve conversations, which allow researchers to discover interviewees' views. This is important since this research seeks to explore people's modeling process that is basically a process of cognitive learning. A major way of studying people's thought is to examine their oral expressions.

The first important issue that needs to be discussed about interviews is sampling. This research adopted a combination of dimensional and snowball sampling. As Brewer and Hunter (2006, 99) claim, this method allows a scholar to combine the survey researcher's concern about representativeness of sampling and the field researcher's concern about gathering "rich" data. The dimensions of sampling in this research are major groups involved in coal mining safety regulation, including government regulators, the industry, unions, and the academy. Within each dimension, samples were not selected randomly. This is because this research needs very detailed information, which can only be known by people who are extensively involved in the process. Therefore, most of those sampled were "elite" individuals who are knowledgeable about the issues that I investigated. I identified some of these individuals simply through reading. For example, I realized that Neil Gunningham is a well-known scholar on systems-based regulation and mining safety through reading literature and included him as a candidate for interview.

I chose others through snowball sampling. Among those interviewees selected through this method, some were sampled atheoretically. In other words, the informant recommended other persons who were worth talking to without any guidance. For instance, during a conversation with David Lauriski, he suggested that I contact Dr. Larry Grayson. Other interviewees were recommended by informants following the direction I gave based on the theoretical framework for this research. For example, expecting that government inspectors (model mongers) and some scholars (model missionaries and mercenaries) would have significant influence on the shift toward risk-based governance in Australia, I asked informants questions such as "who are the most influential government inspectors and scholars you suggest that I talk to?" In addition, some people whom I met in the events that I participated in were selected for interviews. They were selected because I believed that they had extensive knowledge of this field.

Interviews were conducted in a semi-structured style, which integrates the approaches of both formal interviews relying on structured questions and informal conversations. Considering the exploratory nature of my research, I decided that informal conversation has a distinctive advantage because it can allow informants to help me identify key issues that otherwise might not be raised in a more structured interview. However, informal conversation is not efficient. It may take a lot of time and generate a large amount of data, most of which may not be included in the final report. Given time and resource constraints, I could not completely rely on informal conversation. Hence, I also used structured interview questions. These can also increase the validity of the data by providing a chance for cross-checking among different informants.

Nevertheless, I gave different weight to informal conversation based on the stage of the research. I divided interviews into three stages. The first stage included five preliminary interviews, which were undertaken between January and May 2008. In this stage, interviews were very informal because the main purposes of these preliminary interviews were to identify issues for research and to develop the analytical framework. The second stage started when I had a clearer analytical framework. In this stage, I included more structured interview questions, but I still used a fair amount of informal conversation to provide an opportunity to refine my framework. Most interviews were conducted in this stage (August 2008 to January 2009). After January 2009, when I became quite confident about my interpretations, I started the last stage of interviews, in which I used few informal conversations. Often times, I directly discussed my interpretations with respondents. The purposes of the interviews in this stage were to test the validity of my interpretations and to fill gaps in information.

Based on my experience in this research, I believe that informal conversation is necessary for exploratory research because it provides an opportunity to build rapport with informants. Without rapport, one may only receive superficial answers. One way to build rapport is to make interviews reciprocal. In other words, interviewers can use informal conversation to provide interviewees some information and insights that might interest them. As a result, interviewees become not only knowledge givers but also learners. For example, during many interviews, I talked about the nuclear industry's self-regulatory strategies, which could provide lessons for the coal mining industry. Some informants were very interested in what I said. One even asked me to send him relevant information and said that he should interview me.

A total of 41 individuals were interviewed, including 28 Americans and 13 Australians (see Appendix B for the status of interviewees). All interviews were conducted in a one-on-one environment, either face-to-face or on the telephone. Most interviews ranged between 30 and 75 minutes, and some lasted nearly two hours. Moreover, some respondents were interviewed more than once. In particular, three key informants were interviewed more than three times in various circumstances. All interviews except one were taped. My feeling is that all respondents looked comfortable while talking when the voice recorder was on.

The last issue I wish to discuss in this section is how to enhance the validity of interview data. Usually, there are three ways to do this. The first is to cross-check information given by different interviewees. As mentioned earlier, structured interview questions included in my research provided the opportunity to cross-check answers from different people. The second is to conduct repeated interviews with the same informants and cross-check their answers given at different times. In this research, eight informants were interviewed more than once. Some of these repeated interviews were not intended. Rather, I arranged another interview because interviewees did not have enough time to finish all of the questions in the first interview. These seemingly disappointing results turned into an unexpected advantage, providing opportunities to cross-check answers. Another common validity-strengthening strategy is to triangulate data with information gathered from other methods. This is a major advantage of multi-method research (Brewer and Hunter, 2006). In this research, I used interviews as a main source of data and other methods to triangulate the interview data, including document analysis.

Document Analysis

Document analysis is used in almost all qualitative research. The major purposes of document analysis in this research are as follows. The first is to better understand contexts by reviewing scholarly articles, government documents (e.g., mine safety reviews in Australia), government archives (e.g., safety records), and general information on websites. The second is to get detailed information on processes of change, such as MSHA's efforts to move toward the direction of risk-based governance, the process of legislative change in Australia, and the move of the mining industry toward sustainable development. Documents used for this purpose included government archives, conference presentations, industrial reports, speeches, and information on websites. The third purpose is to understand the process of cognitive interpretation. Many documents themselves are actually the outcomes of cognitive interpretation processes. Examples are discussion papers for Australian legislative review and the NIOSH report on major hazard risk assessment projects. Fourth, this research emphasizes individuals' influence on institutions. As such, some remarks and presentations were used as key sources to analyze presenters' thoughts and influence. Finally, as mentioned, document analysis also was used to triangulate the interview data.

Based on this discussion, some limitations of this research have emerged. Since no research design is perfect, it is necessary for a researcher to understand the limitations of a study.

Limitations

This research has four major limitations. The first results from the purpose of the inquiry. As discussed in Chapter 2, the purpose of this study is to fill the gap in existing literature on risk-based governance by examining the concrete mechanisms that lead to the growth of risk-based governance in a particular setting—coal mining safety regulation. Therefore, to what extent the findings of this research can be generalized to other contexts is questionable.

The second limitation involves case selection—the American-Australian comparison. There is a drawback in selecting Australia as a case to compare with the U.S. Unlike Australia where mining safety regulation is only a matter at the state level, both federal and state governments have authority to regulate mining safety in the U.S. In addition, after the passage of the Coal Mine Health and Safety Act (the Coal Act) of 1969, the federal government has become more important than state governments in regulating mining safety. Although a more rigorous research design would have been comparison at the state level, the U.S. case cannot be understood without looking at the dynamics at the federal level. Therefore, the comparison in this research to a large extent is a comparison between federal regulations in the U.S. and state regulations in Australia. This limitation leaves the question of the extent to which the factor of scale influenced the findings of the research. To remedy this limitation, a country with a similar federal regulatory system such as Canada could be chosen. However, this country did not undergo significant regulatory reform as did Australia, although mining legislation in Canada has to some extent moved away from the prescriptive approach. Therefore, Australia is still a better comparable case to explore the research question.

The other two limitations are to some degree related to the problem of accessibility. On the one hand, as discussed earlier, access to the two companies was not deep enough. On the other hand, this research has some bias toward the management perspective. This is partly because the regulatory model discussed in this research—risk-based governance—is what I believe to be a better approach than prescriptive regulation. This model of risk-based governance is considered by some scholars and practitioners who support prescriptive regulation as a pro-business approach. At the same time, most of the interview data were collected from people in the industry, the government, and the academy. Only three informants from unions (two Australians and one American) were included. The reason is that union candidates for interviews were more difficult to access than industry officials were, particularly in the U.S. For example, when I attended a meeting of the West Virginia Board of Coal Mine Health and Safety, I asked Board members representing both the industry and unions for interviews. An industry representative agreed, and all three union representatives declined. They all pointed me to the director of safety or communications at the United Mine Workers of America's (UMWA) national office. It is not the purpose of this study to investigate the reason behind this phenomenon. A tentative explanation based on my observation is that the union's culture is more organized. Union officials are less willing to express their own opinions to the public and tend to follow their official statements.

Appendix B

INTERVIEWEES BY POSITION AND COUNTRY

Status	U.S.	Australia
Management	6	
Safety professionals in companies	2	
Industry associations	2	1
Current or past employees in federal government regulator	5	
Current or past employees in state government regulators	3	4
Congress/parliament	1	
Employees in government advisory bodies		2
Union officials	1	2
Laborers in companies	1	
Researchers in government research institutes	3	
Scholars	2	4
Consultant	1	
Media	1	
Total	28	13

Notes

NOTES TO CHAPTER 1

1. The drawing of the figure borrowed the idea of C. Jennings (2008). The data, however, were collected by the author himself and were recalculated into 2009 dollars based on the Bureau of Labor Statistics Consumer Price Index. ftp.bls.gov/pub/special.requests/cpi/cpiai.txt (accessed February 10, 2011).
2. Before 2005, these elements had been included in many scholarly theories, such as those of the post-regulatory state (Scott, 2004), responsive regulation (Ayers and Braithwaite, 1992), flexible regulation (Rees, 1988), meta-regulation (Parker, 2002), systems-based regulation (Gunningham and Johnstone, 1999), management-based regulation (Coglianese and Lazer, 2003), regulatory society (Braithwaite, 2003), regulatory pluralism (Gunningham and Sinclair, 1999), and smart regulation (Gunningham and Grabosky, 1998).
3. According to Braithwaite (2008, 16–20), security, which stemmed from bonds that created national debts, enabled the replacement of family firms with large corporations based on pooled capital from thousands of shareholders. Moreover, the enforcement of U.S. anti-trust laws actually encouraged mergers because it "tolerated that path to monopoly power while they more effectively outlawed the alternative pathway via cartels and restrictive practices" (p. 19).
4. These internal systems include environmental management, safety and health management, public reporting, and third-party certification systems (Courville, 2004, 210; Levi-Faur, 2005, 21).
5. According to Rogers (2003, 5) "diffusion is the process in which an innovation is communicated through certain channels over time among the members of a social system. It is a special type of communication, in that the messages are concerned with new ideas." According to Dolowitz and Marsh (2000, 5), all studies on lesson-drawing, policy convergence, policy diffusion, and policy transfer "are concerned with a similar process in which knowledge about policies, administrative arrangements, institutions, and ideas in one political setting (past or present) is used in the development of policies, administrative arrangements, institutions, and ideas in another political setting."

   Information and knowledge are distinctive. The former means processed data that are ready to be used. The latter, on the contrary, emphasizes more the psychological process of human beings. Knowledge is the cognitive results obtained from human experience and information. It "resides in one's mind" (Terra and Angeloni, 2009, 2). The two concepts, however, can be used interchangeably because knowledge has to be turned into symbolic information, such as documents, to be transferable. The Webster dictionary also uses these two words to explain each other. For example, information is "knowledge obtained from investigation, study, or instruction," whereas knowledge is "the fact or condition of having information or of being learned." Merriam-Webster 3rd New International Dictionary 1993.

6. According to Rogers (2003, 169), knowledge is the process that occurs when individuals are exposed to an innovation and they try to understand how it works; persuasion means the formation of a favorable or an unfavorable attitude; decision is the process that leads to an adoption or rejection of the innovation; implementation is the process of how adopters put a new idea or practice into use; confirmation is the process of adopters reinforcing or reversing previous decisions.
7. These articles suggest some theoretical frameworks that can be used to analyze the diffusion of regulatory capitalism, including informational networks (Lazer, 2005), policy clustering (Elkins and Simmons, 2005), and policy learning and diffusion (Meseguer, 2005), and they also contribute some case studies of the diffusion of regulatory authorities and some new regulatory strategies in both developed and developing counties.
8. Braithwaite and Drahos's theory of modeling first appeared in Braithwaite's article "A Sociology of Modeling and the Politics of Empowerment" (Braithwaite, 1994).

   I recognize that the theory of modeling shares similarities with some often-used theories of policy process, such as multiple streams and the advocacy coalition framework; they are all sensitive to the cognitive aspect of policy change. A comparison of the theory of modeling with other theoretical frameworks merits another book. Hence, this statement does not preclude the possibility that scholars have applied other theories to address similar questions.
9. This reflects Foucault's idea that power and knowledge are forever connected. As Foucault (1977, 27) notes, there is no power relation without the correlative constitution of a field of knowledge, or any knowledge that does not presuppose and constitute at the same time, power relations.
10. Braithwaite and Drahos (2000, 31) argue that "contemporary capitalism has shifted from industrial to information capitalism. . . . Hegemony in the world system shifted from the control of territory, to industrialization (and the control of capital and labor it implied), to the control of abstract objects."

NOTES TO CHAPTER 2

1. The former stresses the allocation of regulatory resources based upon regulatees' level of risk, whereas the latter emphasizes mandating or encouraging regulated firms to establish internal management systems.
2. The other two research areas that Hutter (2005a, 14) suggests are investigating the extent to which the rhetoric and ideas about risk translate into substantive change and examining how understandings of risk vary between regulators, domains, and countries.
3. Haas (1992, 3) defines epistemic community as "a network of professionals with recognized expertise and competence in a particular domain and an authoritative claim to policy-relevant knowledge within that domain or issue area."
4. After the 1980s, regulatory negotiation or negotiated rulemaking has been encouraged to allow interested parties to get together, negotiate, and reach consensus in the U.S., upon recognition that judicialized rulemaking has led to increasing legal challenges to agency action. However, at the national level, regulatory negotiation has been used only in a limited number of situations for several reasons. First, it is designed to complement, but not replace,

the traditional rulemaking process. Regulatory negotiation is only encouraged, but not required by the Negotiated Rulemaking Act of 1990 (Cooper, 2000, 161). Second, regulatory negotiation is time and resource consuming. It takes a considerable period of time and numerous meetings. Third, there is a strong suspicion that regulatory negotiation may limit participants to a small number of interest groups (Rosenbloom and O'Leary, 1997, 80).

5. This analytical framework does not intend to downplay the influence of political structures, regulatory culture, and industry relations. These structural variables certainly have broader influence. However, since this research takes modeling as a central concept of regulatory change, it emphasizes the role of webs of dialogue as an intervening variable between independent structural variables (political structures, regulatory culture, and industry relations) and the modeling process.

NOTES TO CHAPTER 3

1. Stakeholders include not only those directly affected by mining operations, such as a host of communities, shareholders, and employees, but also many who are indirectly affected by mining operations such as those who are concerned with environmental protection and human rights (MMSD, 2002).
2. This conference aimed to define a future course of action for the sector and prepare reports for the World Summit on Sustainable Development (Rio+10), which would review progress toward sustainable development since 1992. Rio +10 was to be held in Johannesburg, South Africa, in August 2002, three months after the GMI conference.
3. MIRMgate was initially launched in 2004, sponsored by the Minerals Council of Australia (MCA). It is designed to provide carefully selected information on best practices and lessons learned about risk management in the mineral industry.
4. Interview.
5. According to information on its website, Consol Energy has 31 operations in 2007. http://www.consolenergy.com (accessed May 2009).
6. Interview.
7. Interview.
8. Interview.
9. Interview.
10. Interview.
11. The company's name is fictitious to preserve its anonymity.
12. The names are fictitious to preserve their anonymity.
13. Weick (1995) refers to sensemaking as making sense of uncertainties in environments through interaction. According to Weick, Sutcliffe, and Obstfeld (2005), sensemaking starts with chaos. It is an ongoing process in which people retrospectively make plausible sense of what occurs and enact this sense back into the world to make that world more orderly.
14. Interview.
15. The name of the company is fictitious to preserve its anonymity.
16. Interview.
17. Interview.
18. Interview.
19. Interview.
20. The average number of coal mining fatalities per year dropped from 154.7 in 1970–1979, to 94.4 in 1980–1989, to 45.4 in 1990–1999, and to 37.5 in 2000–2008 (MSHA, 2011).

21. Interview.
22. Interview.
23. Interview.
24. Interview.
25. Interview
26. An interviewee said: "the National Mining Association had the foresight to put together the mine safety technology and training commission that they asked me to chair. The association allowed us to be an independent commission with labor representation, company representation, and a lot of experts in different fields, neutral people like academics. They did not interfere at all. They absolutely have a faith here to have this kind of study done. That is when we decided. We embedded in that report that risk management needs to be done. Now they have the foresight to put the group together. The group then makes all these recommendations."
27. Interview.
28. Interview.
29. Interview.
30. According to an informant from NMA, a workshop was scheduled in September 2009, but it was postponed partly because it generated many fewer participants than had been expected.

NOTES TO CHAPTER 4

1. Blacksburg Coal is also a fictitious name.
2. Conversation with participants of Minexpo.
3. According to Coleman (2008) and EIA (1993, 41), the top 20 companies accounted for 82% of production in 2007 as opposed to 55% in 1990.
4. Interview.
5. Interview.
6. For example, on January 26, 2006, the governor of West Virginia signed new mine safety legislation (Senate Bill No. 247) into law. This law requires a quicker response to mine accidents, electronic tracking equipment on miners working underground, and additional portable air stations/supplies in underground mines. This law set the foundation for the Miner Act passed at the federal level almost five months later.
7. According to those speakers from MSHA or NIOSH, there is no true through-earth wireless communications technology available. It has to depend on fixed facilities to some extent.
8. At the time of the interview, the S-Miner Act had been passed by the House and was pending in the Senate. Given the election of Obama, many industry people suspect that the S-Miner Act is very likely to be passed.
9. Interview.
10. Interview.
11. Interview.
12. Interview.
13. Interview.
14. Interview.
15. Interview.
16. The behavior-based approach to safety can be traced back to H. W. Heinrich, who published *Industrial Accident Prevention: A Scientific Approach*, in 1931. He was the first scholar who indicated that as many as 95% of all workplace accidents are caused by unsafe acts.
17. This example is given in an anonymous article Behavior-Based Safety Evolves; http://hr.blr.com/whitepapers.aspx?id=18907 (accessed July 2009).

18. Interview.
19. Interview.
20. Interview.
21. Interview.
22. A possible way to evaluate the influence of risk management is to focus on spending on significant citations.
23. Interview.
24. Interview.
25. Interview.
26. According to the General Duty Clause of the OSHA Act, employers are liable only if the hazards, which are not covered by specific standards, are recognized by employers.
27. Interview.
28. The company's name is fictitious.
29. Interview.
30. Interview.
31. Cards are available at different areas around the mine. The dropping-in box is put at a place where there is not much visibility because some people may be embarrassed if they are seen dropping a card in it.
32. This research emphasizes the influence of people's cognition and their perceived properties of the models. It does not totally deny that American coal mining companies prefer BBS because this model costs less to adopt. However, the cost to adopt a model is related to the social cognitive factors discussed earlier. The less complex the model is perceived to be and the more expertise available to put it in place, the less the costs of adopting the model. One may also question whether the different results of adopting BBS rather than risk management in terms of improving safety performance make BBS more favorable to the American coal mining industry. Based on my observation, the industry has not been fully aware of the difference between BBS and risk management, let alone their different results. I have not seen any research comparing the different results of the two models. This also reflects an important concept in decision-making—bounded rationality (Simon, 1985). Decisions are made without complete information. Oftentimes, it is results that are observed, but not the actual products of a particular model, that lead to the adoption of one model over the other.

NOTES TO CHAPTER 5

1. Interview.
2. This is because, according to section 28 of the OSH Act 2000 (NSW), the only way for employers to defend themselves is to prove that, first, it was not "reasonably practicable" to comply with the provisions of the Act; or, second, the commission of the offense was due to causes over which the person had no control and it was impracticable for the person to make provision against the happening (Gunningham, 2007, 21).
3. According to Gunningham (2007, 21), however, it is unfortunate that the Queensland legislation does not include general duties because risk management implies that employers can postpone taking action on lower ranking risks.
4. According to Gunningham (2007, 23), judges in the NSW Industrial Relation Commission and Workers Compensation Tribunal interpreted that the requirements of employers' general duties in the OHS Act 2000 (NSW) implied the need of risk management systems. One judge says: "[The employer must take] a structured, systemic approach to safety in everything which is

touched by the operations of the defendants. It is not enough to endeavor to comply with these obligations on an ad hoc basis looking at particular matters from time to time . . . employers are required to actively assess and take account of all risks that might foreseeably arise. Systems need to be created to deal with these risks and to the extent possible, eliminate them. Employees need to be instructed and trained to apply these systems. The employer needs to assess from time to time whether those systems are working and whether employees are following them. This involves supervision."

5. This argument does not deny that regulatory reform of Australian coal mining safety is also a functional response in dealing with risk. According to an informant, governments certainly do not want problems; they do not want "issues on the headlines of papers." Whenever there is a disaster, government tends to rethink whether a regulatory framework works well or not.
6. Interview.
7. Interview.
8. Interview.
9. Interview.
10. Interview.
11. According an informant, the introduction of a principal hazards management plan is primarily the result of unions' unwillingness to let the regulation be adopted without some checks and balances. This is because the unions believe that the cause of the Moura accident was the poor management of risk management systems at the mine site. He said that "the new act would be much more process oriented and much less prescriptive than they are now, had there not been the Moura disaster in 1994."
12. Interview.
13. Interview.
14. Interview.
15. Interview.
16. Interview.
17. Interview.
18. Interview.
19. Interview.
20. Interview.
21. Interview.
22. Interview.
23. The three companies were prosecuted following fatal accidents in the late 1990s. Then they appealed their convictions to the Court of Appeal. According to Ellison (2005), two main grounds for the appeal are exactly those last two concerns of mine operators. They claim that the provision of the OHS law under which the mine managers and surveyors were convicted is too broad, and they argued that the OHS law inappropriately shifted the burden of proof in prosecutions to the employer. According to the OHS Act, the only thing that prosecutors need to do is to demonstrate that a breach occurred. They do not need to prove employer negligence. It is the employers who must establish a defense to escape liability. The three companies argued that the reversal of the traditional burden of proof is not consistent with the presumption of innocence.
24. Interview.
25. Interview.
26. Interview.
27. Interview.
28. Interview.
29. Interview.

30. Interview.
31. In an interview, a senior government inspector mentioned that two key individuals who worked in the metalliferous group have been important figures in the nation. One has become a very senior official at Xstrata, one of the biggest mining companies in the world. The other one, whose "father was the most powerful union person in Queensland and one of the most in Australia," has now been elected to the Australian Senate.
32. Interview.
33. According to information published in *Mine Safety News* by the NSW Department of Mineral Resources (DMR), the MSAC before 2006 was chaired by Dennis Else, a professor at the University of Ballarat. The members included three representatives of government from the Department of Mineral Resources: Graham Terrey (Director of the Mine Safety and Environment Division, DMR) and two Assistant Directors of the Mine Safety and Environment Division, DMR—Neville Sneddon and Rod Morrison; three industry representatives: Bob Humphris (NSW Minerals Council) and John Vale (Minerals Council of Australia), and Tony Browne (Crushed Stone & Sand Association); and two union representatives: Huge McDermott (Australian Workers Union) and Ron Land (CFMEU) (DMR 1999).
34. Interview.
35. Interview.
36. Interview.
37. Interview; compulsory voting was gradually abolished in Australia after the 1980s.
38. Interview.
39. Interview.
40. In the movie *Monty Python and the Holy Grail,* King Arthur fights with Black Knight, who is not prepared to call it a draw until losing all four limbs. This scene reminded Maher of the never-say-die attitude of the mining companies and unions (Maher, 2006, 4).

    In addition, the more harmonious relationship between the industry and unions does not mean there is no conflict at all. In fact, the conservative federal government (1996–2007) had undermined the role of unions and tripartite collaboration (Saksvik and Quinlan, 2003; Gunningham, 2008). For example, according to Gunningham (2008, 341), the Workplace Relations Act of 1996 led to the decline of collective bargaining; the introduction of Australian Workplace Agreements (AWAs) increased the role of individual contracts. The tripartite National Occupational Health and Safety Commission was substantially downsized (Saksvik and Quinlan, 2003, 41).
41. Interview.
42. Interview.
43. Interview.
44. Interview.

NOTES TO CHAPTER 6

1. For example, the Federal Coal Mine Health and Safety Act of 1941 authorized federal inspectors to enter and inspect health and safety hazards in all types of coal mines, but did not give them authority to establish and enforce standards. Public Law 80–328 in 1947 granted authority to the government agency to establish safety standards for bituminous coal and lignite mines without granting enforcement power.
2. Interview.

3. Interview.
4. For example, a fatality occurred in the Brooks Run Mining Company's Mercer Deep Mine in 2004. A continuous mining machine operator was pinned between the ripperhead of the continuous mining machine and the coal rib of a crosscut. The report identified two causal factors. One was the failure to comply with the approved roof control plan, which requires persons to be in a safe location from the continuous mining machine while tramming in remote mode. The report also showed the correct action on this causal factor. The management initiated a more stringent policy regarding the proximity of personnel to continuous mining machines. This policy requires that "no person shall be positioned between the continuous mining machine and the coal ribs when the continuous miner pump motors are enabled" and "no person shall position themselves within two rows of roof bolts in front of the continuous mining machine cutting head" (MSHA, 2004).
5. They are agreements with American Society of Safety Engineers, the Association of Equipment Manufactures, the Bituminous Coal Operator's Association and the National Mining Association, the Gypsum Association, the International Association of Bridge, Structural, Ornamental and Reinforcing Iron Workers, the Industrial Minerals Association—North America, the International Union of Operating Engineers, the National Safety Council, the National Stone, Sand and Gravel Association, and the Portland Cement Association.
6. According to ICF (2003), in the late 1990s, an internal Coal Reinvention Work Group and the Metal and Non-Metal's (MNM) Voluntary Compliance Partnership (VCP) were set up to examine possibilities for increasing inspection efficiency. The Coal Reinvention Work Group put forward several recommendations such as modifying the requirement for reviewing the Uniform Mine File (UMF) before an inspection and prioritizing the areas for inspection. MNM proposed the VCP initiative. Neither the Coal Reinvention Work Group recommendations nor the MNM's VCP program was put in place.
7. Interview.
8. Interview.
9. For example, the *Handbook* requires inspectors to conduct step-by-step inspections as follows: MSHA personnel should proceed to the area selected for inspection as quickly as possible after arriving at the mine site. When inspection personnel travel to a working section while conducting inspections or investigations, they should check all the working places for imminent dangers as soon as practical after arrival on the section and before examining equipment or observing any cycle of operation. Even if mine management has, at the time of arrival, started to take action to correct the condition(s) or practice(s), cite the violations (ICF, 2003, 19).
10. Interview.
11. Interview.
12. Interview.
13. During the period of 1979 to 1984, some employers in the construction industry that already had excellent safety records were invited to join California's CCP program. According to Rees (1988), these companies agreed to establish labor-management safety committees to carry out OSHA's regulatory responsibilities. In return, the companies were exempted from routine inspections.

    OSHA's VPP started in 1982. Like the California CCP, this program targeted companies with "exemplary safety and health records" (Needleman, 2000, 74). The selected companies must have safety and health management systems in place, have plans to identify and control hazards, and audit their safety and health practice in a continuous manner. Again, the participants would be removed from the list of routine inspections. In addition, the best VPP participants would be awarded a logo "Star," "representing outstanding work-sites considered models of excellence" (Needleman, 2000, 75).

OSHA's national CCP program was announced in 1997 as a major initiative of the Clinton administration's "New OSHA" strategy. According to DTF (2000, 79), *The New OSHA: Reinventing Worker Safety and Health* put forward three sets of regulatory reform initiatives. One was to initiate a paradigm shift in OSHA from command and control to providing employers a choice between partnership and a traditional enforcement relationship. If a firm had effective health and safety management systems, OSHA offered partnership. Another strategy focused on the most serious hazards and the most dangerous workplaces. OSHA's national CCP initiative was consistent with these two strategies. According to Needleman (2000, 79), OSHA identified around 12,500 employers with unusually high lost workday injury and illness rates based on survey data. OSHA put these employers on its primary inspection list, while providing most of them an alternative. If they chose to join the CCP and voluntarily develop safety and health plans, they would be removed from the primary inspection list.

14. Interview.
15. Interview.
16. Interview.
17. Interview.
18. Interview.
19. Vicarious consequence means "the consequence one feels he would get based on observing the consequence of others' similar action."
20. Interview.
21. Interview.
22. Interview.
23. Interview.
24. Interview.
25. In Australia, specific standards are put in codes of practice and guidelines, not in statutes and regulations. However, as discussed previously, coal mine operators have to adopt codes of practice or guidelines to demonstrate duty of care. Otherwise, they have to show evidence of why they adopted alternatives.
26. Interview.
27. An informant said: "We were living in the time of continuing success. Nobody thought the system needed to be changed. The trends were going in the right directions. The fatality numbers were dropping, and the injury and illness numbers were dropping. It was a time that there was maybe a false sense of security where the mindset was we are doing it right, something is working here. And then Sago occurred; it was a wake-up call for the entire industry."
28. An interviewed union official said: "We are in favor of risk management of potential risk . . . in what you referred to as physical area. In terms of dealing with people's behavior, the more training miners can get, the better safety. As long as it is not used as a tool for something other than safety, I think risk management probably is a good thing."
29. Interviews.
30. Interview
31. Interview.
32. Some informants acknowledged that there is a lack of partnership on regulatory policy issues in the U.S. generally, although some partnerships do exist

when conducting research on particular technical matters, such as dust and noise control, mainly through the platform of NIOSH.

33. Right-to-work laws are enforced in 22 states in the U.S. They prohibit agreements between unions and employers making union membership a condition of employment.
34. Interview.
35. Interview.
36. Interview.
37. This example also illustrates the importance of scholars from outside the industry. This is because insiders are usually constrained by their existing cognitive structure, whereas outsiders can often bring people to think "outside the box." As this example shows, these innovative solutions that have been undertaken in Utah were recommended by the UMSC, chaired by a law professor at the University of Utah.
38. Interview.
39. In fact, tripartism is also subject to criticism in Australia because it means anyone apart from trade unions, companies, and governments would not be included. Ayres and Braithwaite (1992, 58) argue as well that a "simple tripartism model would provide too narrow a basis for public interest group (PIG) participation" in rule-making processes. Indeed, having recognized the shortcoming of tripartism, the latest Mine Safety Advisory Council in NSW has included independent members. Therefore, the MSAC now has actually become a multipartite, rather than a tripartite body (Wran and McClelland, 2005b).
40. Interview.
41. Interview.

NOTES TO CHAPTER 7

1. Nodes of governance often emerge when a state or non-state actor network partners when they lack the capacity, resource, and knowledge to manage events (Braithwaite 2008, 99). A node of governance is an institutional site that brings together different governance authorities and provides ways of thinking to manage a course of events (Burris, Drahos, and Shearing, 2005, 37; Wood and Shearing, 2007, 13).
2. In a prescriptive and adversarial regulatory regime, controlling government power is the dominant purpose of the law. Administrative agencies are seen simply as neutral instruments to implement the will of the legislature, whose members are elected by the people.

NOTES TO APPENDIX A

1. Internal validity refers to "the truthfulness of the claim that one variable causes another" (Dooley, 2001, 163). External validity addresses the question of generalizability (Campbell & Stanley, 1963, 5).
2. I do not release the companies' names for confidentiality reasons.

Bibliography

Acil. (1997) *Review of Mine Safety in NSW, A Report to the Minister for Mineral Resources and Fisheries*, March 14, 1997, Sydney.

Anderson, E. (2006) "The Epistemology of Democracy," Episteme*: A Journal of Social* Epistemology 3(1), 8–22.

Ayers, I., & Braithwaite, J. (1992) *Responsive Regulation: Transcending the Regulation Debate*. New York; Oxford: Oxford University Press.

Bancroft, R. (2000) *Development of the Queensland Coal Mining Safety and Health Regulation, 2000*. Paper presented to Queensland Mining Industry Health and Safety Conference, 2000. http://www.qrc.org.au/conference/_dbase_upl/2000_spk002_bancroft.pdf (accessed August 16, 2009).

Bandura, A. (1986) *Social Foundations of Thought and Action: A Social Cognitive Theory*. Englewood Cliffs, NJ: Prentice-Hall.

Bassey, M. (1999) *Case Study Research in Educational Settings: Doing Qualitative Research in Educational Settings*. Buckingham/Philadelphia: Open University Press.

Bean, R. (1989) *International Labor Statistics: A Handbook, Guide, and Recent Trends*. London; New York: Routledge.

Beck, U. (1992) *Risk Society: Toward a New Modernity*. Beverly Hills, CA: Sage.

Bennear, L. (2007) "Are Management-Based Regulations Effective? Evidence from State Pollution Prevention Programs." *Journal of Policy Analysis and Management,* 26(2), 327–348.

Billingham, R. (2000) *Mining and Quarrying Safety and Health Regulation, 2000*. Paper presented to Queensland Mining Industry Health and Safety Conference, 2000. http://www.qrc.org.au/conference/_dbase_upl/2000_spk004_billingham.pdf (accessed August 16, 2009).

Black, J. (2005) "The Emergence of Risk-based Regulation and the New Public Risk Management in the United Kingdom" *Public Law* (Autumn), 512–548.

———. (2010) "Risk-based Regulation: Choices, Practices and Lessons Being Learnt." In *OECD Reviews of Regulatory Reform, Risk and Regulatory Policy: Improving the Governance of Risk*, OECD Publishing, Paris, 185–236.

Box, R. (2002) "Pragmatic Discourse and Administrative Legitimacy." *American Review of Public Administration* 32(1), 20–39.

Braithwaite, J. (1993) "Responsive Regulation for Australia" In P. Grabovsky & J. Braithwaite (Eds.) (1993), *Business Regulation and Australia's Future*. Canberra: AIC, 81–96.

———. (1994) "A Sociology of Modeling and the Politics of Empowerment." *British Journal of Sociology* 45, 443–479.

———. (2003) *Meta-Regulation for Access to Justice*. Paper presented at the General Aspects of Law (GALA) Seminar Series, University of California, Berkeley, November 13.

———. (2008) *Regulatory Capitalism: How It Works, Ideas for Making It Work Better.* Cheltenham: Edward Elgar.

Braithwaite, J., & Drahos, P. (2000) *Global Business Regulation.* Cambridge: Cambridge University Press.

Brewer, J., & Hunter, A. (2006) *Foundations of Multimethod Research: Synthesizing Styles.* Thousand Oaks, CA: Sage.

Burke, K. (1945) *A Grammar of Motives.* New York: Prentice-Hall.

Burris, S., Drahos, P., & Shearing, C. (2005) "Nodal Governance." *Australian Journal of Legal Philosophy,* 30, 31–58.

Campbell, D. T., & Stanley, J. C. (1963) *Experimental and Quasi-Experimental Designs for Research.* Chicago: Rand McNally & Company.

Clark, L. (1999) "The Politics of Regulation: A Comparative-Historical Study of Occupational Health and Safety Regulation in Australia and the United States." *Australian Journal of Public Administration* 58(2), 94–104.

Coglianese, C. (2010) "Management-Based Regulation: Implications for Public Policy." In *OECD Reviews of Regulatory Reform, Risk and Regulatory Policy: Improving the Governance of Risk.* OECD Publishing, Paris, 159–184.

Coglianese, C., & Lazer, D. (2003) "Management-Based Regulation: Prescribing Private Management to Achieve Public Goals." *Law & Society Review* 37(4), 691–730.

Coglianese, C., & Nash, J. (2001) *Regulating from the Inside: Can Environmental Management Systems Achieve Policy Goals?* Washington, DC: Resources for the Future.

Coleman, L. (2008) National Mining Association 2007 Coal Producer Survey. http://www.nma.org/pdf/members/coal_producer_survey2007.pdf (accessed July 25, 2009).

Cooper, P. (2000) *Public Law and Public Administration* (3rd ed.). Itasca, IL: F. E. Peacock Publishers,

Courville, S. (2004) "Making Sense of Corporate Responsibility Tools" In Galea, C. (Ed.), *Teaching Business Sustainability*: Vol. 1. *From Theory to Practice,* 210–217. Sheffield, UK: Greenleaf.

DMR. (1999) *Mine Safety News.* Department of Mineral Resources, NSW. http://www.dpi.nsw.gov.au/__data/assets/pdf_file/0008/80945/NSW-Mine-Safety-Update-1999-March.pdf (accessed August 16, 2009).

DNRME. (2004) *Mine Inspectorate Safety Newsletter.* www.dme.qld.gov.au/zone_files/inspectorate . . . /newsletter_aug04.pdf (accessed July 28, 2009).

Dobbin, F., Simmons, B., & Garrett, G. (2007) "The Global Diffusion of Public Policies: Social, Construction, Coercion, Competition, or Learning?" *Annual Review of Sociology* 33, 449–472.

DOL. (2011) *FY2011 Congressional Budget Justification MSHA. www.dol.gov/dol/budget/2011/PDF/CBJ-2011-V2–12.pdf* (accessed February 7, 2011).

Dolowitz, D., & Marsh, D. (2000) "Learning from Abroad: The Role of Policy Transfer in Contemporary Policy-Making." *Governance* 13(1), 5–23.

Dooley, D. (2001) *Social Research Methods* (4th ed.). Upper Saddle River, NJ: Prentice Hall.

DTF. (2000) *Health and Safety Review: Going Forward: Appendices.* Department of Treasury and Finance, Victoria, Australia

Duncan, H. D. (1969) *Symbols and Social Theory.* New York: Oxford University Press.

Dye, D. (2005) *Statement of David G. Dye Acting Assistant Secretary of Labor for Mine Safety and Health before the Subcommittee on Labor, Health and Human Services and Education Committee on Appropriations, U.S. House of Representatives.* http://www.msha.gov/media/congress/2005/04072005.htm (accessed July 29, 2009).

EIA. (2009a) *Coal Production and Number of Mines by State and Mine Type.* http://www.eia.doe.gov/cneaf/coal/page/acr/table1.html (accessed August 16, 2009).

———. (2009b) *Coal Supply and Demand: Exports and Imports.* http://www.eia.doe.gov/cneaf/coal/page/special/exports_imports.html (accessed August 15, 2009).

———. (2009c) *Country Analysis Briefs: Australia Coal, Energy Information Administration.* http://www.eia.doe.gov/emeu/cabs/Australia/Coal.html (accessed July 25, 2009).

———. (2009d) *U.S. Coal Supply and Demand: 2008 Review, Energy Information Administration.* http://www.eia.doe.gov/cneaf/coal/page/special/feature.html (accessed July 25, 2009).

———. (1993) *The Changing Structure of the U.S. Coal Industry: An Update, Energy Information Administration.* http://tonto.eia.doe.gov/ftproot/coal/051393.pdf (accessed July 25, 2009).

Elkins, Z. & Simmons, B. (2005) "On Waves, Clusters, and Diffusion: A Conceptual Framework." *ANNALS of the American Academy of Political and Social Science*, 598, 33–51.

Ellison, M. (2005) *Mine Health and Safety: Could you be Liable?* http://www.hg.org/articles/article_675.html (accessed August 15, 2009).

Foucault, M. (1977) *Discipline and Punish: the Birth of the Prison.* London: Penguin Books.

Frick, K., Jensen, P., Quinlan, M. & Wilthagen, T. (Eds.). (2000) *Systematic Occupational Health and Safety Management: Perspectives on an International Development.* Amsterdam: Pergamon.

Frick, K., & Wren, J. (2000) "Reviewing Occupational Health and Safety Management—Multiple Roots, Diverse Perspectives, and Ambiguous Outcomes." In Frick, K., Jensen, P., Quinlan, M., & Wilthagen, T. (Eds.), (2000) *Systematic Occupational Health and Safety Management: Perspectives on an International Development.* Amsterdam: Pergamon, 17–42.

Gallick, J. (2008) *Accident Reduction.* Presentations to Bluefield Coal Symposium, Bluefield, West Virginia, August 2008.

GAO. (2007) *MSHA's Revised Hiring Process Has Improved the Agency's Recruiting Efforts, but Its Human Capital Strategic Plan Does Not Adequately Project or Address Its Future Workforce Needs.* http://www.gao.gov/htext/d07704r.html (accessed July 31, 2009).

Geller, S. (1994) "Ten Principles for Achieving a Total Safety Culture." *Professional Safety* 39(9), 18–24.

———. (1997) "What Is Behavior-Based Safety, Anyway?" *Occupational Health and Safety* 66(1), 25–35.

Gioia, D. A., & Chittipeddi, K. (1991) "Sensemaking and Sensegiving in Strategic Change Initiation." *Strategic Management Journal* 12, 433–448.

Grayson, R. L., & Watzman, B. (2001) "History and Overview of Mine Health and Safety." In Karmis, M. (Ed.), *Mine Health and Safety Management.* Littleton, CO: Society for Mining, Metallurgy, and Exploration, 1–16.

Gunningham, N. (1999) *The Development of New Regulatory Model for Occupational Health and Safety in the New South Wales Coal Industry: An Issues Paper for the New South Wales Department of Mineral Resources.*

———. (2004) *Evaluating Mine Safety Legislation in Queensland.* www.ohs.anu.edu.au/publications/pdf/wp%2042%20-%20Gunningham.pdf (accessed February 5, 2011).

———. (2006) *Safety Regulation and the Mining Inspectorate—Lessons from Western Australia. http://hdl.handle.net/1885/43146* (accessed February 5, 2011).

———. (2007) *Mine Safety: Law, Regulation, Policy.* Sydney: The Federation Press.

———. (2008) "Occupational Health and Safety, Worker Participation and the Mining Industry in a Changing World of Work." *Economic and Industrial Democracy*, 29(3), 336–361.
Gunningham, N., & Grabosky, P. N. (1998) *Smart Regulation: Designing Environmental Policy*. Oxford: Clarendon.
Gunningham, N., & Johnstone, R. (1999) *Regulating Workplace Safety: System and Sanctions*, Oxford: Oxford University Press.
———. (2000) "The Legal Construction of OHS Management Systems." In Frick, K., Jensen, P., Quinlan, M., & Wilthagen, T. (Eds.), *Systematic Occupational Health and Safety Management: Perspectives on an International Development*. Amsterdam: Pergamon, 125–146.
Gunningham, N., & Sinclair, D. (1999) "Regulatory Pluralism: Designing Policy Mixes for Environmental Protection." *Law & Policy* 21, 49–76.
Haas, P. M. (1992) "Introduction: Epistemic Communities and International Policy Coordination." *International Organization*, 46(1), 1–35.
Harvey, B. (2007) *Keynote Address—Utah Mining Association, 92nd Annual Meeting*. Park City, Utah, August 23, 2007. http://www.consolenergy.com/Newsroom/Speech5.aspx (accessed August 16, 2009).
Heinrich, H. W. (1931) *Industrial Accident* Prevention. New York: McGraw-Hill.
Holt, J. (2008) *Zero Accident*. Presentations at SME, Pennsylvania State University, March 19, 2008.
Hood, C., James, O., Jones, G., Scott, C. & Travers, T. (1999) *Regulation inside Government*. Oxford: Oxford University Press.
Hopkins, A. (1999) *Managing Major Hazards: The Lessons of the Moura Mine Disaster*. St. Leonards, N.S.W.: Allen & Unwin Pty LTD.
———. (2007) Beyond Compliance Monitoring: New Strategies for Safety Regulators. *Law & Policy*, 29(2), 210–225.
Hutter, B. (2001) "Is Enforced Self-Regulation a Form of Risk-Taking? The Case of Railway Health and Safety." *International Journal of the Sociology of Law*, 29, 379–400.
———. (2005a) *The Attractions of Risk-Based Regulation: Accounting for the Emergence of Risk Ideas in Regulation*. www.lse.ac.uk/collections/CARR/pdf/Disspaper33.pdf (last accessed April 2008).
———. (2005b) "Risk Management and Governance." In Eliadis, P., Hill, M., & Howlett, M. (Eds.), *Designing Governments: From Instruments to Governance*. London: McGill-Queen's University Press, 303–321.
Iannacchione, A., Varley, F., & Brady, T. (2008) *The Application of Major Hazard Risk Assessment (MHRA) to Eliminate Multiple Fatality Occurrences in the U.S. Minerals Industry*, http://www.cdc.gov/niosh/mining/pubs/pubreference/outputid2859.htm (accessed May 16, 2009),
ICF. (2003) *Mine Inspection Program Evaluation*. Final Report to U.S. Department of Labor Mine Safety and Health Administration, September 30, 2003. www.msha.gov/Readroom/FOIA/SpecialReports/ICFInspectionReport/MSHAMineInspectionProgramEvaluationReport.doc (accessed May 15, 2009).
ICMM. (2009) *Sustainable Development Framework: 10 Principles*. http://www.icmm.com/our-work/sustainable-development-framework/10-principles (accessed August 14, 2009).
Jennings, C. (2008) *Coal Mine Safety Shortchanged by Years of Budget Cuts. http://www.ombwatch.org/node/8588* (accessed February 17, 2011).
Jennings, N. (2009) *The NSW Mine Safety Advisory Council: Addressing OHS Fundamentals Together*. http://www.dpi.nsw.gov.au/__data/assets/pdf_file/0004/148495/MSAC—-Addressing-OHS-fundamentals-together.pdf (accessed July 25, 2009).

Joy, J. (2004) "Occupational Safety Risk Management in Australian Mining." *Occupational Medicine*, 54, 311–315.

———. (2008) *Risk Assessment and the U.S. Mining Industry*. Presentation to Minexpo, Las Vegas, September 22–24, 2008. http://www.minexpo.com/Presentations/Joy.pdf (accessed August 16, 2009).

Joy, J., & Griffiths, D. (2007) *National Minerals Industry Safety and Health Risk Assessment Guideline*. http://www.mishc.uq.edu.au/Files_for_download/NMISHRAG/NMISHRAG_v6.pdf (accessed August 16, 2009).

Kaarbo, J., & Beasley, R. K. (1999) "A Practical Guide to the Comparative Case Study Method in Political Psychology." *Political Psychology* 20(2), 369–391.

Kagan, R. (1991) "Adversarial Legalism and American Government." *Journal of Policy Analysis and Management* 10(3), 369–406.

Kelman, S. (1981) Regulating *America*, Regulating Sweden*: A Comparative Study of Occupational Safety and Health Policy*. Cambridge, MA: MIT Press.

Kingdon, J. (1984) *Agendas, Alternatives, and Public Policies*. Boston: Little Brown.

Krause, T. (1997) "Trends and Developments in Behavior-based Safety." *Professional Safety* 42(10), 20–25.

———. (2001) "Moving to the 2nd Generation in Behavior-Based Safety." *Professional Safety* 46(5), 27–32.

Lauriski, D. (2004) *Remarks at the Safety Achievement Awards Luncheon, Indiana Mineral Aggregates Association, Indianapolis, Indiana, February 5, 2004*. http://www.msha.gov/MEDIA/SPEECHES/2004/02052004.htm (accessed August 16, 2009).

Lazer, D. (2005) "Regulatory Capitalism as a Networked Order: the International System as an Informational Network." *ANNALS of the American Academy of Political and Social Science*, 598, 52–66.

Levi-Faur, D. (2005) "The Global Diffusion of Regulatory Capitalism." *ANNALS of the American Academy of Political and Social Science*, 598, 12–32.

———. (2008) "Forward." In Braithwaite, J. (2008) *Regulatory Capitalism: How It Works, Ideas for Making It Work Better*, Cheltenham: Edward Elgar, vii–x.

Levi-Faur, D., & Jordana, J. (2005) "The Making of a New Regulatory Order." *ANNALS of the American Academy of Political and Social Science*, 598,6–9.

Lijphart, A. (1971) "The Comparable-Cases Strategy in Comparative Research." *Comparative Political Studies*, 8, 158–177.

Lobel, O. (2004) "The Renew Deal: The Fall of Regulation and the Rise of Governance in Contemporary Legal Thought." *Minnesota Law Review*, 89, 342–470.

Maher, T. (2006) *Co-operation Is Vital: Presentation to NSW Minerals Industry, OHS Conference*. www.cfmeu.com.au/storage/documents/TMPaper-OHS2006b.pdf (accessed July 28, 2009).

Maldonado-Maldonado, A. (2004) *An Epistemic Community and Its Intellectual Networks: The Field of Higher Education in Mexico*. PhD dissertation, Boston College.

Markley, R. (2004) *A Conversation with Dave Lauriski: On the Success of His Administration*. http://rockproducts.com/mag/rock_conversation_dave_lauriski/ (accessed July 29, 2009).

Marshall, C., & Rossman, G. (1999*) Designing Qualitative Research* (3rd ed.). Thousand Oaks, CA: Sage.

McSwite, O. C. (1997) *Legitimacy in Public Administration: A Discourse Analysis*. Thousand Oaks, CA: Sage.

Mead, G. H. (1950) *Mind, Self, and Society*. Chicago: University of Chicago Press.

Meseguer, C. (2005) “Policy Learning, Policy Diffusion, and the Making of a New Order.” *ANNALS of the American Academy of Political and Social Science*, 598, 67–82.

Minerals Council of Australia. (1999) *Safety and Health Performance Report, 1998–1999. http://www.minerals.org.au/__data/assets/pdf_file/0009/4878/1998_99SPR.pdf* (accessed February 8, 2011).

———. (2007) *Safety and Health Performance Report, 2006–2007. http://www.minerals.org.au/safety/safety_and_health_performance/index.html* (accessed February 8, 2011).

MMSD. (2002) *Towards Change: The Work and Results of MMSD North America/ Mining, Minerals and Sustainable Development North America.* http://www.iisd.org/pdf/2002/mmsd_mining_toward_change.pdf (accessed July 24, 2009).

MSHA. (2004) *MSHA Coal Mine Safety and Health Report of Investigation, Underground Coal Mine, Fatal Machinery Accident, Mercer Deep Mine, Brooks Run Mining Company, I.D. No. 46–08875.* http://www.msha.gov/FATALS/2004/ftl04c07.pdf (accessed August 16, 2009).

———. (2008a) *Preliminary Regulatory Economic Analysis for Safety Standards Regarding the Recommendations of the Technical Study Panel on the Utilization of Belt Air and the Composition and Fire Retardant Properties of Belt Materials in Underground Coal Mining.* http://www.msha.gov/regs/rea/e8–13631beltair.pdf (accessed August 14, 2009).

———. (2008b) *SLAM Risk: Coal and Metal/Nonmetal Fatalities, July–December 2007*, MSHA/National Mine Health and Safety Academy training materials.

———. (2009a) *MSHA Enforcement Data* (obtained through information request).

———. (2009b) *MSHA Digital Library: Accident Report Search* http://arllib.msha.gov/awweb/main.jsp (accessed July 2009).

———. (2010a) *Data Retrieval System: Upper Big Branch Mine—South. http://www.msha.gov/drs/ASP/BasicMineInfonew.asp* (accessed December 10, 2010).

———. (2010b) *Mine Safety and Health at a Glance. http://www.msha.gov/MSHAINFO/FactSheets/MSHAFCT10.HTM (accessed February 7, 2011).*

———. (2010c) *Mining Industry Accident, Injuries, Employment and Production Statistics: All Coal Mining Data.* http://www.msha.gov/ACCINJ/BOTHCL.HTM (accessed February 10, 2011).

———. (2011) *Coal Fatalities for 1900 through 2010.* http://www.msha.gov/stats/centurystats/coalstats.asp (accessed February 10, 2011).

MSTTC. (2006) *Improving Mine Safety Technology and Training: Establishing U.S. Global Leadership.* www.eme.psu.edu/Faculty/grayson/msttc_report.pdf (accessed April 2008).

Needleman, C. (2000) “OSHA at the Crossroads: Conflicting Frameworks for Regulating OHS in the United States.” In Frick, K., Jensen, P., Quinlan, M., & Wilthagen, T. (Eds.), *Systematic Occupational Health and Safety Management: Perspectives on an International Development.* Amsterdam: Pergamon, 67–86.

NMA. (2009a) *Economic Contribution of US Mining in 2007.* National Mining Association. http://www.nma.org/pdf/pubs/mining_economic_report.pdf (accessed July 25, 2009).

———. (2009b) *Summary of the Mine Improvement and New Emergency Response (MINER) Act of 2006 Requirements.* http://www.nma.org/topic_pages/mine_safety.asp (accessed July 27, 2009).

Ober, J. (2008) *Democracy and Knowledge: Innovation and Learning in Classical Athens.* Princeton: Princeton University Press.

OECD. (2009) *OECD Labor Force Statistics, Annual Force Statistics, Summary Table: Trade Union Density.* http://stats.oecd.org/Index.aspx?DatasetCode=ALFS_SUMTAB (accessed July 26, 2009).

OPM. (2009) *Qualification Standards for General Schedule Positions: Mine Safety and Health Inspection Series*. http://www.opm.gov/qualifications/standards/IORs/gs1800/1822.htm (accessed July 29, 2009).

Parker, C. (2002) *The Open Corporation: Effective Self-regulation and Democracy*, Cambridge; New York: Cambridge University Press.

Pettit, P. (1997) *Republicanism*. Oxford: Clarendon.

Poplin, G. S., Miller, H. B., Ranger-Moore, J., Bofinger, C. M., Kurzius-Spencer, M. , Harris, R. B. and Burgess, J. L. (2008) "International Evaluation of Injury Rates in Coal Mining: A Comparison of Risk and Compliance-Based Regulatory Approaches." *Safety Science*, 46(8), 1196–1204.

Powell, W., & DiMaggio, P. (1991) *The New Institutionalism in Organizational Analysis*. Chicago: University of Chicago Press.

Power, M. (2004) *The Risk Management of Everything*. London: Demos.

———. (2007) *Organized Uncertainty: Designing a World of Risk Management*. Oxford: Oxford University Press.

Przeworski, A., & Teune, H. (1970) *The Logic of Comparative Social Inquiry*. New York: Wiley-Interscience.

Rees, J. (1988) *Reforming the Workplace: A Study of Self-regulation in Occupational Safety*. Philadelphia: University of Pennsylvania Press.

———. (1994) *Hostages of Each Other: The Transformation of Nuclear Safety since Three Mile Island*. Chicago: University Of Chicago Press.

Robson, L. S., Clarke, J. A., Cullen, K., Bielecky, A., Severin, C., Bigelow, P. L., Irvin, E., Culyer, A., Mahood, Q. (2007) "The Effectiveness of Occupational Health and Safety Management System Interventions: A Systematic Review." *Safety Science*, 45, 329–353.

Rogers, E. (2003) *Diffusion of Innovations* (5th ed.). New York: Free Press.

Rose, R. (1991) "What Is Lesson-Drawing?" *Journal of Public Policy* 11(1), 3–30.

Rosenbloom, D., & O'Leary, R. (1997) *Public Administration and Law* (2nd ed.). New York: M. Dekker.

Sabatier, P., & Jenkins-Smith, H. (1988) "Special Issue: Policy Change and Policy Oriented Learning: Exploring an Advocacy Coalition Framework." *Policy Science* 21(2/3), 123–277.

Saksvik, P., & Quinlan, M. (2003) "Regulating Systematic Occupational Health and Safety Management: Comparing the Norwegian and Australian Experience." *Industrial Relations*, 58(1), 33–59.

Scott, C. (2004) "Regulation in the Age of Governance: The Rise of the Post-Regulatory State." In Jordana, J. & Levi-Faur, D. (Eds.), *The Politics Of Regulation: Institutions and Regulatory Reforms for the Age of Governance*. Cheltenham: Edward Elgar, 145–174.

Sheehy, B., & Dickie, P. (2002) *Facing the Future: MMSD Australia*. http://www.goodpracticemining.org/documents/jon2/100_mmsdaustralia.pdf (accessed July 24, 2009).

Simon, H. (1985) "Human Nature in Politics: The Dialogue of Psychology with Political Science," *American Political Science Review* 79: 293–304.

Starke, L. (2003) *Breaking New Ground: Mining, Minerals and Sustainable Development*. London: Earthscan Publications.

Terra, J., & Angeloni, T. (2009) *Understanding the Difference between Information Management and Knowledge Management*. www.providersedge.com/ . . . /Understanding_the_Difference_Between_IM_and_KM.pdf (accessed September 25, 2009).

Turner, V. (1974) *Dramas, Fields, and Metaphors*. Ithaca: Cornell University Press.

UMSC. (2008) *Utah Mine Safety Commission: Report and Recommendations to Governor Jon M. Huntsman, Jr.* http://minesafetycommission.utah.gov/press/UMSC%20Report%20Final.pdf (accessed August 16, 2009)

UMWA. (2007) *UMWA's Roberts Says MSHA, Coal Operators Not Fulfilling Mine Safety Responsibilities.* http://mineworkers.org/index.php?q=node/82 (accessed July 29, 2009).

Vogel, D. (1986) *National Styles of Regulation: Environmental Policy in Great Britain and the United States.* Ithaca: Cornell University Press.

Walker, J., & Howard, S. (2002) *Finding the Way Forward: How Could Voluntary Action Move Mining towards Sustainable Development?* http://www.goodpracticemining.org/documents/jon2/finding_the_way.pdf (accessed July 24, 2009).

Weick, K. (1995) *Sensemaking in Organizations.* Thousand Oaks, CA: Sage.

Weick, K., Sutcliffe, K. M., & Obstfeld, D. (2005) "Organizing and the Process of Sensemaking." *Organization Science,* 16, 409–421.

Ward, K. (2009) *How the Bush Administration Reversed Decades of Progress on Mine Safety.* http://www.washingtonmonthly.com/features/2007/0703.ward.html (accessed July 30, 2009).

Wood, J., & Shearing, C. (2007) *Imagining Security.* Cullompton: Willan Publishing.

World Coal Institute. (2009a) *Coal Market and Transportation.* http://www.worldcoal.org/coal/market-amp-transportation/ (accessed August 14, 2009).

———. (2009b) *Coal Mining.* http://www.worldcoal.org/coal/coal-mining/ (accessed August 14, 2009).

Wran, H. N., & McClelland, J. (2005a) *NSW Mine Safety Review: Report to the Hon Kerry Hickey, MP, Minister for Mineral Resources, Appendices.* http://www.dpi.nsw.gov.au/minerals/safety/mine-safety-initiatives/wran-mine-safety-review (accessed July 29, 2009).

———. (2005b) *NSW Mine Safety Review: Report to the Hon Kerry Hickey, MP, Minister for Mineral Resources, Full Report.* http://www.dpi.nsw.gov.au/minerals/safety/mine-safety-initiatives/wran-mine-safety-review (accessed July 29, 2009).

Yin, R. (2009) *Case Study Research: Design and Methods* (4th ed.). Thousand Oaks, CA: Sage.

Index